Besseboua Omar

Manual of analysis methods and techniques

Besseboua Omar

Manual of analysis methods and techniques

ScienciaScripts

Imprint
Any brand names and product names mentioned in this book are subject to trademark, brand or patent protection and are trademarks or registered trademarks of their respective holders. The use of brand names, product names, common names, trade names, product descriptions etc. even without a particular marking in this work is in no way to be construed to mean that such names may be regarded as unrestricted in respect of trademark and brand protection legislation and could thus be used by anyone.

Cover image: www.ingimage.com

This book is a translation from the original published under ISBN 978-620-6-70437-9.

Publisher:
Sciencia Scripts
is a trademark of
Dodo Books Indian Ocean Ltd. and OmniScriptum S.R.L publishing group

120 High Road, East Finchley, London, N2 9ED, United Kingdom
Str. Armeneasca 28/1, office 1, Chisinau MD-2012, Republic of Moldova, Europe
Managing Directors: Ieva Konstantinova, Victoria Ursu
info@omniscriptum.com

Printed at: see last page
ISBN: 978-620-8-39394-6

Contents

Foreword

This book is the culmination of years of teaching on analysis methods and techniques, aimed at second-year Master's students in animal production and nutrition in the Department of Agronomic Sciences and Biotechnology. It aims to support the learning of different biological, biochemical and immuno-enzymatic analysis techniques and methods. The aim is to introduce students to laboratory work, particularly as part of their final dissertation.

The manuscript is divided into four main chapters arranged in chronological order of analysis methods, as follows:

1- Spectral methods.
2- Fractionation methods.
3- Marking methods.
4- Electron microscopy.

Like all work, it can contain errors and shortcomings. So it's always encouraging and motivating to receive corrections, advice and recommendations from fellow teachers and researchers working in the field.

Chapter 1

Spectral methods.

3.1 Molecular absorption spectrometry

3.1.1 UV-Visible absorption spectroscopy

Spectrophotometry is a technique for measuring how light interacts with materials. When light strikes a material, it can be reflected, transmitted, scattered or absorbed. In addition, the material may emit light at a different frequency due to the energy it acquires from the incident light (such as electroluminescence) or due to its temperature (such as incandescence) [Germer TA, et al., 2014]. Various types of spectroscopy and spectrophotometry are well-known and widely used techniques for the identification and quantification of substances in research, as well as in industrial and chemical laboratories. For example, in chemistry and pharmacy, UV-visible spectrophotometry is a fundamental method for analysing samples using the Beer-Lambert-Bouguer law. In biochemistry and molecular biology, spectrophotometric analysis is essential for determining the concentration of biomolecules in a solution, such as DNA, RNA or proteins [Trumbo TA, et al., 2013]. In clinical laboratories, manual and automated spectrophotometry is widely used to analyse blood, urine and other body fluids [Rand RN. 1972].

3.1.1.1 Absorption principle :

The law of absorption is the basic principle of UV-visible spectrophotometry. This law deals with the relationship between the thickness of the absorbing material and the concentration of the sample solution, known as Beer-Lambert's law or simply Beer's law. This law states that the amount of light absorbed is proportional to the concentration of the absorbing substance and the thickness of the absorbing material [Upadhyay A, et al., 2009] (Figure 1).

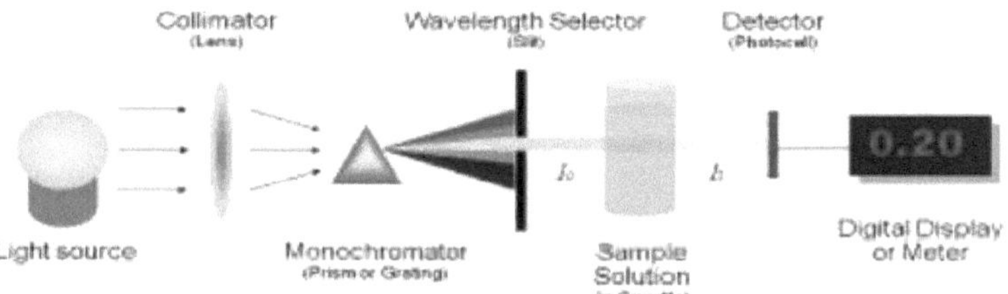

Figure 1: Basic instrumentation of a spectrophotometer [https://chem.libretexts.org].

3.1.1.2 Instrumentation

The UV-visible spectrophotometer consists of a light source, sample holders, a monochromator and a detector [De Caro CA, Claudia H. 2015].

3.1.1.2.1 Light source :

Hydrogen lamps and deuterium lamps are used as UV light sources, while the tungsten filament lamp is the most commonly used for visible light.

3.1.1.2.2 Sample holders :

In the UV and visible ranges, cuvettes are used as sample holders, made of quartz or ordinary glass. Generally, in the UV region, quartz or silica cells are used, while in the visible region, glass cells are used, and these cuvettes generally have a standard path length of 1 cm.

3.1.1.2.3 Monochromators :

A monochromator converts polychromatic radiation into monochromatic radiation, producing very narrow wavelength bands.

3.1.1.2.4 Detectors :

Photovoltaic cells, phototubes and photomultipliers are commonly used as detectors in the UV and visible range [Upadhyay A, et al., 2009]. The following block diagram (Figure 2) shows the main parts of the UV-visible spectrophotometer [Upadhyay A, et al., 2009].

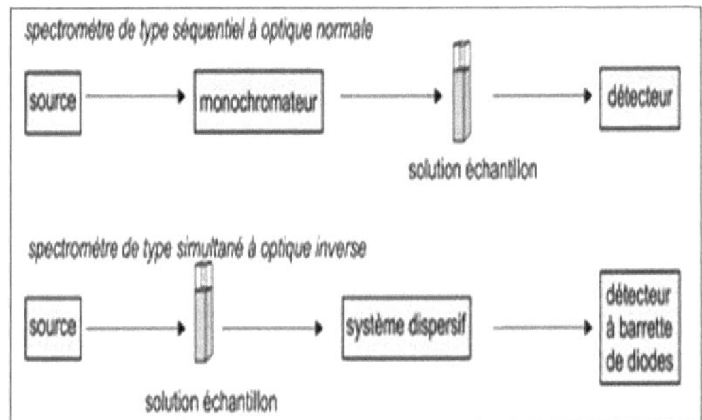

Figure 2. Block diagram of the UV-visible spectrophotometer. (Francis Rouessac, Annick Rouessac, 2004, P 151)

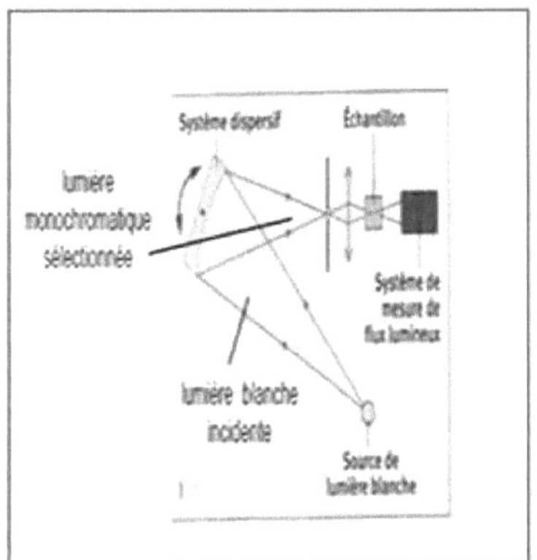

Diagram 1 of the principle of a spectrophotometer

UV and visible absorption spectroscopy is a very common method in laboratories. It is based on the property of molecules to absorb light of a specific wavelength (Figure 3).

The UV-visible range extends from around 800 to 10 nm.

- visible: 800 nm (red) - 400 nm (indigo)
- near-UV: 400 nm - 200 nm

- UV-far: 200 nm - 10 nm.

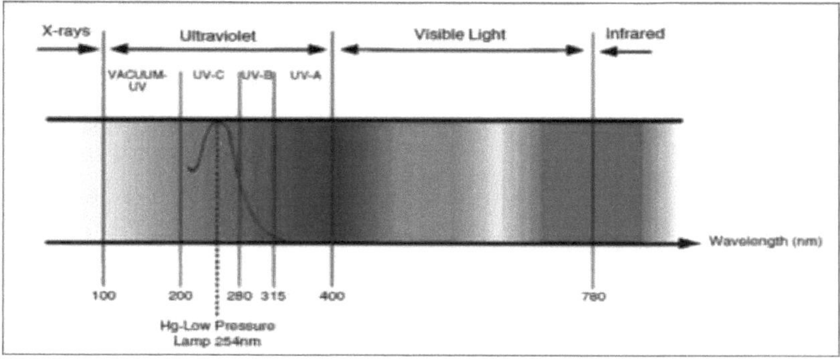

Figure 3: Electromagnetic Spectrum.

3.1.1.3 Beer - Lambert's law

Absorbance is a function of solute concentration, as shown by Beer-Lambert's law: $A = \log (Io/I) = \varepsilon . \lambda . C$

- A = absorbance without unit
- Io = incident light intensity (before interaction with the solute)
- I = transmitted light intensity
- ε = extinction coefficient (which depends on the wavelength) :

1. If the concentration of the solute is in M (or $mol.L^{-1}$), ε is in $M^{-1}.cm^{-1}$, this is the molar extinction coefficient: ε_M
2. If the concentration of the solute is in % (mass/volume), ε is in $g^{-1}.L.cm^{-1}$, this is the weight extinction coefficient: $\varepsilon_{1\%}$.

- λ = otic path length (in cm)
- C = concentration of the solute (the unit depends on that of the extinction coefficient)

3.1.1.4 Protocol for measuring absorbance using a spectrophotometer

To measure the absorbance **A** of a coloured substance in aqueous solution for a wavelength **λo** :

- a cell containing the reference solution is placed in the spectrophotometer (the zero point of the spectrophotometer A = 0 is set by a "blank")
- a cuvette containing a solution of the coloured substance to be analysed is placed in the spectrophotometer.
- we read the value of **A**.

NB: Before each absorbance measurement, the zero must be checked. The blank used must contain the solution from which the element under study has been removed.

The absorbance **A** of a solution of a chemical species generally depends on :

a- its concentration **C** in the solution ;

b- the wavelength **λ** of the incident radiation: **A = ^λ)** [A(λ) = 0 means that the substance passed through is not absorbed or is transparent to radiation of wavelength λ]. c- the thickness ***e*** of solution passed through (the greater *e*, the greater A and vice versa);

d- the nature of the absorbent substance.

3.1.1.5 Applications of UV-visible spectrophotometry

3.1.1.5.1 Pharmaceutical analysis :

UV-visible spectrophotometry is a widely used technique for determining drug concentration in pharmaceutical analysis. For example, this technique is used in the determination of

etravirine in raw materials and pharmaceutical formulations. The spectrum of etravirine is shown below. It is an antiviral drug that shows a maximum absorption at 414 nm (in the visible range) when reacted with NaOH and 1,2-naphthoquinone-4-sulfonate.
Details are shown in **Figure 4** [Murali D, et al., 2014].

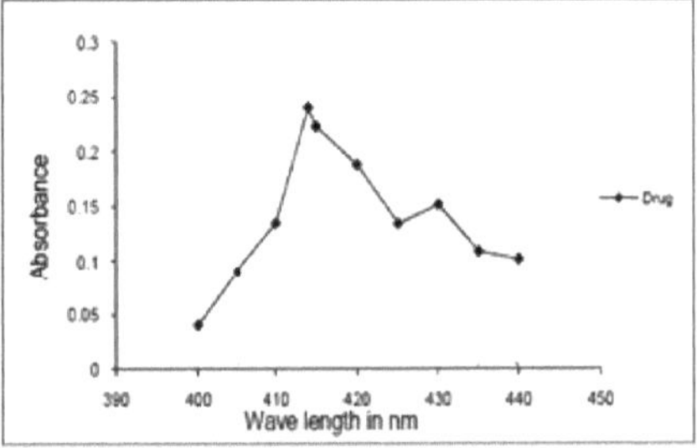

Figure 4: UV-visible spectrum of etravirine.

Quantitative analysis using UV-visible spectrometry is widely used (much more so than qualitative analysis) thanks to the use of Beer-Lambert's law.

Applications include :

a- Determination of iron in water or in a drug.

b- Dosing of active molecules in a pharmaceutical preparation.

c- Determination of benzene in cyclohexane.

3.1.1.5.2 Vaporisation studies of low-volatility compounds :

The UV-visible spectrophotometer is used for the determination of vaporisation of low volatile compounds in the vapour state located above the sample in the condensed state [Verevkin SP, et al., 2018].

UV-visible spectroscopy is also used for: a- the identification of pure analytes that are not subject to decomposition, in particular for the identification of nucleic acids. This study has the advantage of identifying new genetic material discovered in various microbes and other species. **Figure 5** shows the absorption of DNA in UV light [Ling Li X, et al., 2013].

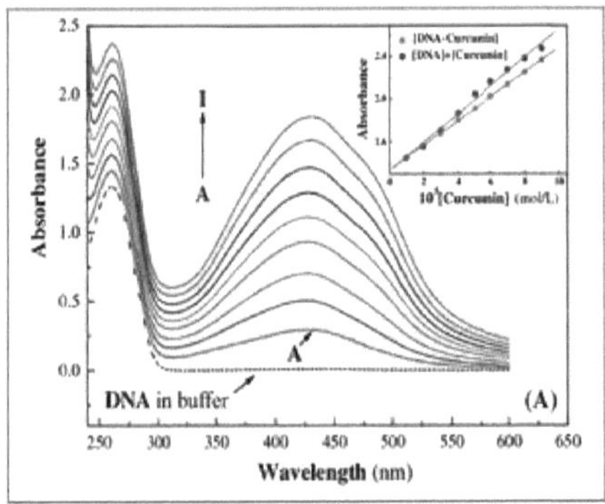

Figure 5: UV absorption spectrum of DNA.

Other applications include quality control, monitoring reaction kinetics, determining acid dissociation constants or complexation constants, determining molar masses, etc.

In molecular biology, it is used during DNA extraction to quantify the DNA and determine its purity. The wavelength used is 260 nm, which is the zone of maximum absorbance of nucleic acids. A second measurement at 280 nm is used to check the purity of the extraction, i.e. the presence of residual proteins in the DNA solution.

b-The quantification and identification of organic compounds was carried out using the UV-visible spectroscopy technique. This technique is very useful for the analysis of new drugs developed in the pharmaceutical industry [Passos MLC, et al. 2019].

UV-visible spectrophotometry is a widely used technique in biochemistry for determining the micromolar concentrations of substances in blood, urine and other body fluids. It is also used for both species determination and the study of biochemical processes [Rojas FS, CanoPavón JM. 2005].

UV-visible spectroscopy is also used to assess the colour index of insulating oil in transformers **(Figure 6)** [Leong YS, et al., 2018].

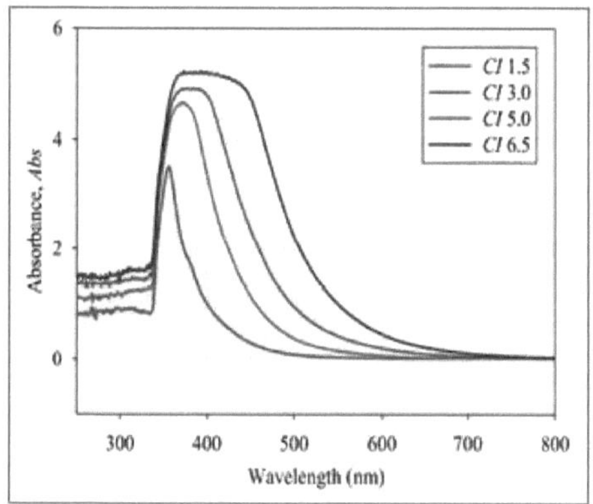

Figure 6. Optical absorbance spectra of four oil samples with different colour indices (CI), based on ASTM D 1500, after the application of an ND filter [Leong YS, et al., 2018].

A case study was conducted combining UV-visible spectroscopy and chemometrics to determine the interaction between human serum albumin (ASH) and gold nanoparticles (AuNPs). Data from UV-visible spectroscopy and chemometrics concerning the interaction of the protein (ASH) with the nanoparticles (AuNPs) were used to establish the thermodynamic, kinctic and structural parameters in order to understand the evolution of the interaction between the protein and the nanoparticles **(Figure 7)** [Yong Wang YN. 2014].

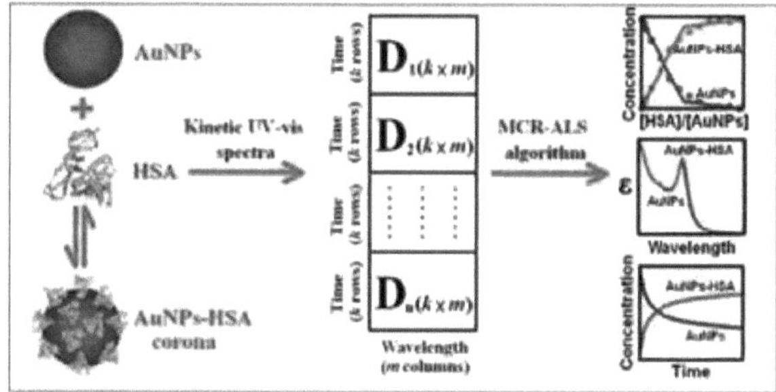

Figure 7. interaction of human serum albumin (HSA) with citrate-coated gold nanoparticles (AuNPs) [Yong Wang YN. 2014].

3.2 Spectrofluorometry

3.2.1 Principle

The phenomenon whereby a molecule, after absorbing radiation, emits radiation of a longer wavelength is known as fluorescence. When a compound absorbs radiation, it passes to an excited level, then returns to the fundamental level, either in a single step by emitting radiation of the same wavelength as it absorbed, or progressively by emitting quantas of radiation corresponding to each energetic step with a longer wavelength. This phenomenon leads to the formation of fluorescence spectra. Fluorescence is an extremely brief phenomenon (10^{-7} s or less) and can therefore provide information on events occurring in less than 10^{-7} s. Fluorometry also follows

Beer-Lambert's operating principle.

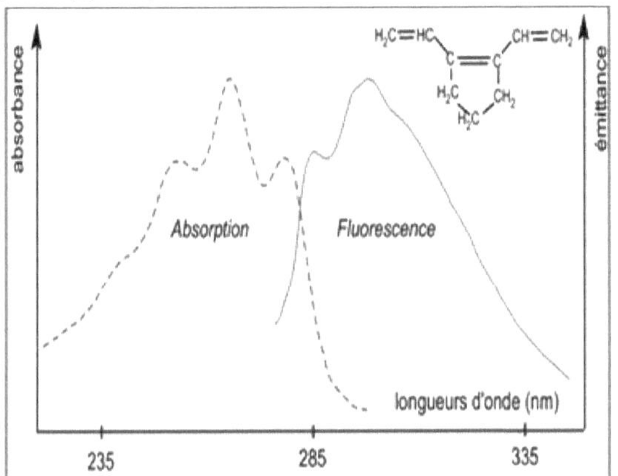

Figure 8 Representation of absorption spectra on the same graph. (Francis Rouessac, Annick Rouessac, 2004, P 206)

Figure 8 gives an example of the apparent mirror symmetry of the absorbance and fluorescence spectra of many compounds. To obtain this representation, the absorbance and emittance spectra are combined on the same graph with a double scale (Francis Rouessac, Annick Rouessac, 2004, P 206).

Fluorometry is an important analytical tool for determining extremely low concentrations of substances that fluoresce.

3.2.2 Instrumentation

The instrumentation of a spectrofluorometer differs from that of a spectrophotometer in two important respects, in addition to a few minor variations (**Figure 9**).

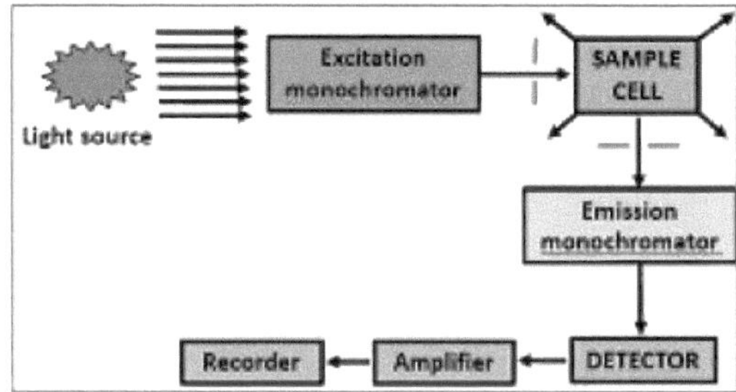

Figure 9: Spectrofluorometer instrumentation.

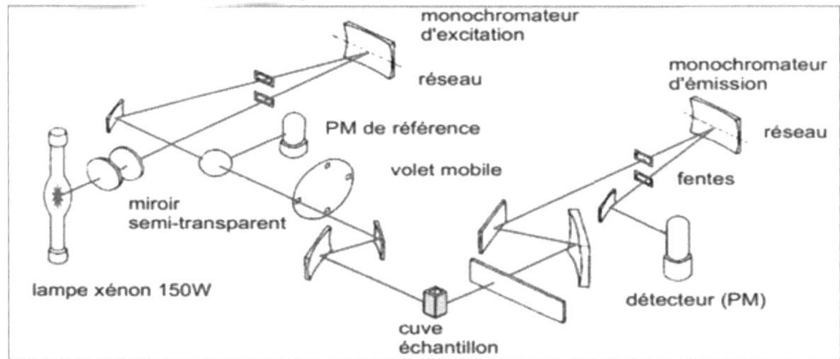

Figure 10 Diagram of the Shimadzu F-4500 spectrofluorometer.

A fraction of the incident beam, reflected by a semi-transparent mirror, strikes a reference photodiode. The signals from the two detectors are compared to eliminate source drift. This single-beam process provides the stability typical of dual-beam devices. Spectra often show small differences when they come from different devices (reproduced with permission from Shimadzu). (Francis Rouessac, Annick Rouessac, 2004, P 214)

a. In a spectrofluorometer, there are two monochromators instead of one, as in a spectrophotometer. One monochromator is placed before the sample holder, and another after it.

b. As fluorescence is at its strongest between 25 and 30°C, the sample holder is fitted with a device to maintain the temperature.

The main components of a spectrofluorometer, shown in **Figure 10**, are as follows:

3.2.3 Applications

a. Identifying the 3D structure of proteins: proteins are made up of an assembly of 20 amino acids. For drug discovery using computational methods, it is very important to know the correlation between the three-dimensional structure and the functions of proteins. This information cannot always be obtained successfully by X-ray crystallography and electron microscopy. Spectrofluorometry can therefore provide reliable information about the three-dimensional structure of proteins while preserving their native structure.

b. In the food sector: fluorescence spectroscopy plays an important role in the determination of numerous food components, adulterants, additives and contaminants.

c. Quality control in food processing: fluorescence spectroscopy is a rapid and sensitive analytical method for characterising food products. For example, some authors have recently studied the impact of heat treatment on vitamin A in milk samples using fluorescence spectra.

The fluorescence spectra show that heat treatment resulted in a decrease in fluorescence intensity at both 320 and 290 nm in milk samples milk samples heated at 75°C for 10 minutes, compared to milk samples heated at 55°C for the same period. Details are shown in **(Figure 11)** [Karoui R. 2016].

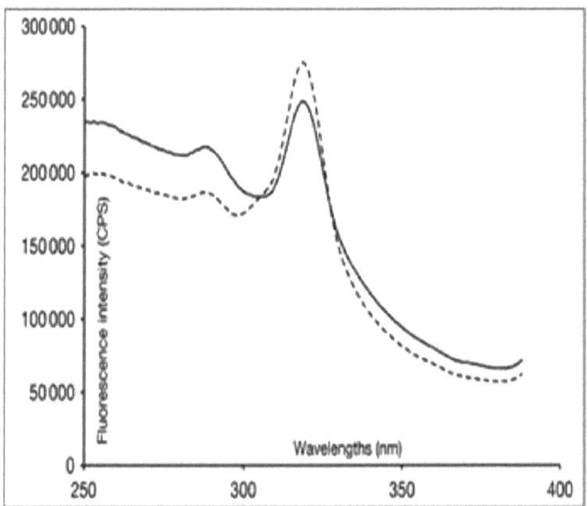

Figure 11. Changes in the fluorescence spectra of vitamin A obtained on milk samples heated at 55°C and 75°C for 10 minutes.

d. The most common applications for the spectrofluorometer include qualitative analysis, quantitative analysis (including the determination of riboflavin, thiamine, hormones such as cortisol, restrogens, serotonin and dopamine, organophosphate pesticides, carcinogens in tobacco smoke, drugs such as lysergic acid and barbiturates, porphyrins, cholesterol and even certain metal ions) ; as well as studies on the structure of proteins (**proteins containing FAD**).

In biochemistry, fluorescence has many applications for quantifying proteins or nucleic acids using reagents that bind specifically to these compounds.

Current classic applications of fluorescence include the measurement of polycyclic aromatic hydrocarbons in drinking water using HPLC. This detection method is well suited to achieving the very low thresholds imposed by legislation. The same process can also be used to measure aflatoxins and many other organic compounds (adrenaline, steroids, vitamins).

Food industry: assays of proteins, amines, amino acids, vitamins, additives, pesticide residues, microtoxins, etc.

Pharmaceutical industry, biology: assays Anaesthetics, analgesics, neuroleptics, tranquillisers, diuretics, sulphonamides, antibiotics .

Microfluorometry: analysis at cellular level Flow cytofluorometry: cell sorting

Chemical industry Aromatic compounds, aldehydes, ketones, petroleum products, rubber, dyes .

3.3 Atomic absorption spectrophotometry (AAS)

3.3.1 Atomic absorption and flame emission

Atomic absorption spectrometry (AAS) and flame emission spectrometry (FE), also known as flame photometry, can be used to measure virtually any type of sample, selected from a list of around 70. For some elements, their sensitivity means that concentrations of less than mg/L can be achieved (Francis Rouessac, Annick Rouessac, 2004, p 241).

3.3.2 Bohr's rule

In the particular case of atomic absorption, we work with free atoms: these atoms can absorb photons and thus enter excited states. As the quantity of photons absorbed is proportional to the number of atoms of the absorbing element, absorption can be used to measure the concentrations of the elements that we have decided to measure.

3.3.3 Principle

When the molecules in the sample are volatilised, the atoms produced absorb certain wavelengths of light from a source that produces an atomic spectrum characteristic of the molecule.

3.3.4 Instrumentation for SAA

3.3.4.1 The basic components of an atomic absorption spectrophotometer are [Butcher DJ. 2005] (figure 12 and 13):

a. Atomisers: the atomisers most commonly used in SAA are electrothermal atomisers (**ETAs**). They are used to convert sample molecules into individual gaseous atoms capable of absorbing light from the source.

b. Light source: generally, a hollow cathode lamp is used as the light source in AAS to produce a certain wavelength of light that is ideally absorbed by the gaseous atoms. Nowadays, instruments with single- and double-beam optics are available.

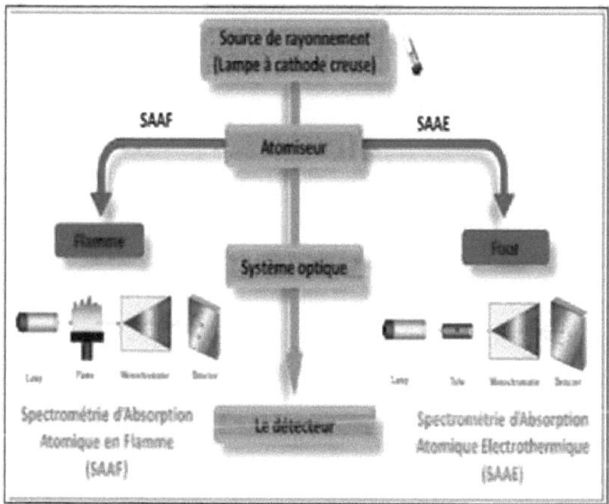

Figure 12: Atomic absorption spectrometry equipment

c. Isolation and quantification of the wavelengths of interest can be achieved using detectors, and to control the operation of the instrument, as well as to collect and process the data, the computer system is a great help.

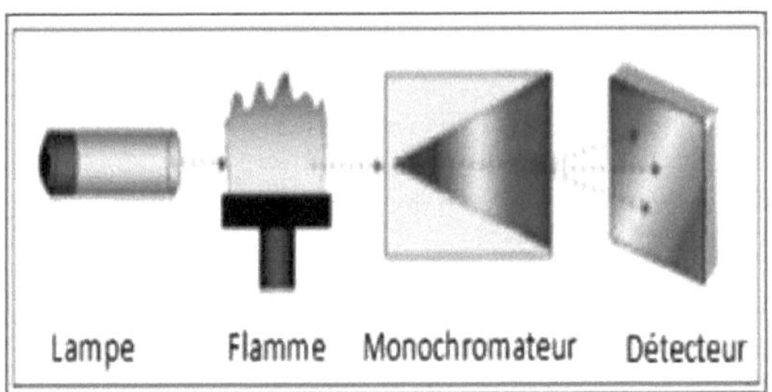

Figure 13: Atomic absorption spectrometry equipment in

3.3.5 Temperature programme

The oven is heated according to a specific programme for the analyte in 04 steps **(Figure 14).**

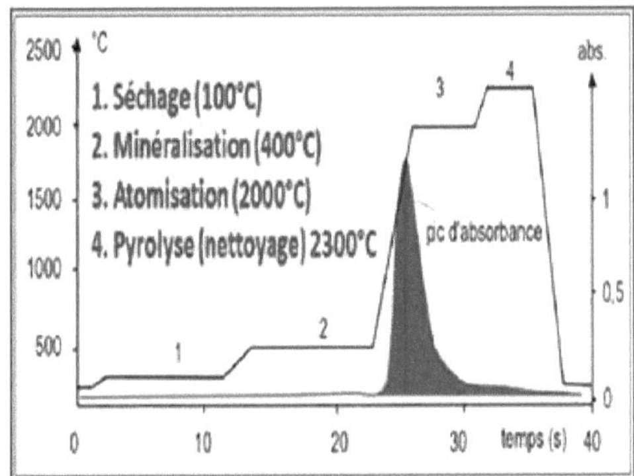

Figure 14: Temperature programme for atomic absorption spectrometry electrothermal

3.3.6 Applications

a. SAA is a sensitive and highly selective spectrometric technique for the determination of many elements at trace and ultra-trace levels, such as Cd, Cr, Zn, Cu, Ag, Mn, Mg, Hg, As, Sc, etc., at picogram level in soils, sediments and plant samples. For example, this technique is very useful for scientists working on research into the impact of mining and industrial activities.

b. Atomic absorption spectrometry (AAS) can be used to estimate metals in body fluids such as serum and whole blood, in addition to urine and tissues, as part of toxicological investigations in clinical studies [Calatayud JM, Icardo MC. 2005].

c. SAA has proved to be a very useful tool, particularly for determining food quality. In this respect, it is involved in the determination of alkali and alkaline earth metals, as these elements act as micro-constituents in food samples.

d. SAA is one of the most useful techniques for checking water quality.

3.4 Atomic emission spectroscopy

Atomic emission spectroscopy is a spectrochemical technique that enables elements to be analysed quantitatively, within a detection limit that varies between parts per million and parts per billion. In theory, all elements except argon can be detected. This method of analysis has the advantage of being able to measure the concentration of several elements in a sample simultaneously, being effective over a wide range of concentrations, which is not possible with absorption spectroscopy. In addition, the sample can be analysed directly, whether in solid, liquid or gaseous form. Analysis by atomic emission spectrometry is a general method for measuring elements, based on the optical study of radiation emitted by atoms that have passed into an excited, generally ionised, state.

3.4.1 Operating principle

The atoms are thermally excited in the flame or plasma so that they re-emit their spectrum. By studying the spectra detected, we can see which elements make up the sample (Figure 15).

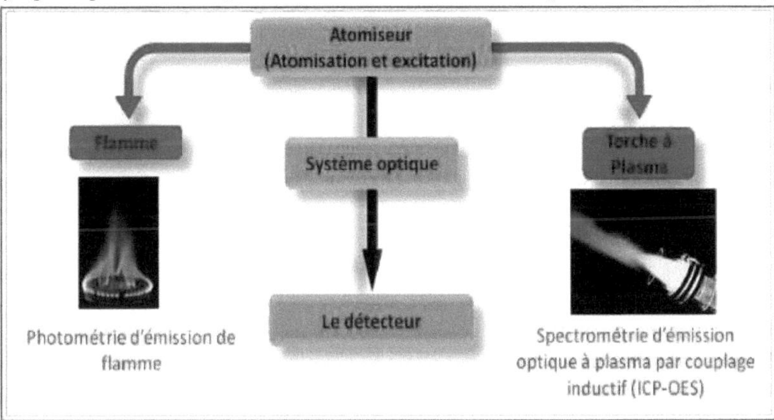

Figure 15: Atomic emission spectrometry equipment

3.4.2 Application

Atomic emission techniques are widely used and offer a number of advantages over absorption spectroscopy:

a- certain elements can be analysed with greater sensitivity and less interference

b- Atomic emission allows qualitative analyses to be carried out, which is not the case with absorption. In emission spectroscopy, it is the sample itself that is the source of light. This means that several elements can be analysed simultaneously.

3.5 Nuclear magnetic resonance (NMR) spectroscopy

3.5.1 Principle

Nuclear magnetic resonance (NMR) is one of the most useful and powerful techniques for determining molecular structure. The principle of NMR is based on the fact that when a strong magnetic field and a radio frequency transmitter are applied to the molecules in the sample, the atomic nuclei of these molecules are excited and form spectral lines in the spectrum [Günther H. 2013].

3.5.2 Instrumentation

The components of an NMR spectrometer are as follows:

a. Radiation source: a radiofrequency (RF) transmitter which generates the radiofrequency current. This current is delivered to the transmitter coil, which creates a signal used to excite the protons in the magnetic field.

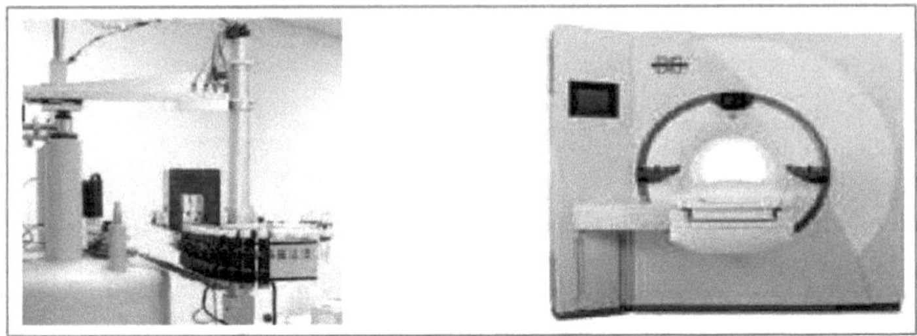

Figure 16 Samples and magnets for NMR equipment.

On the left, robotised introduction of a sample in solution, placed in a "tube of NMR", within the magnetic field produced by a superconducting coil maintained at the temperature of liquid helium (reproduced with the permission of Bruker); on the right, a large electromagnet designed for the introduction of a very specific type of sample: the human body (part of a magnetic resonance imaging machine).

(Francis Rouessac, Annick Rouessac, 2004, P 307)

b. A superconducting magnet, which produces a magnetic field in the central volume of the magnet. The magnetic field produced varies from 1 to 10 teslas. Advanced NMR spectrometers are equipped with superconducting solenoids that generate a magnetic field in excess of 3.5 T. The magnet includes a space to hold the sample probe, as well as a set of room temperature coils (RTS) to reduce the inhomogeneity of the magnetic field in the active volume of the sample.

c. A receiver receives the absorbed signal in digitised form, generally referred to as "free induction decay" (FID).

d. A computer, an amplifier and an analogue-to-digital converter (ADC).

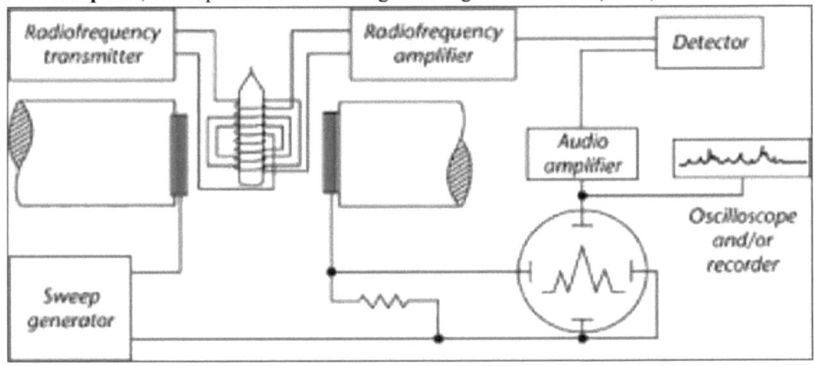

Figure 17. Components of the NMR spectrometer

The NMR spectrum results from the absorption by the sample of some of the frequencies sent by this electromagnetic source. Interpretation of the signals (position, appearance, intensity) provides a range of information about the sample, which is all the more easily interpreted if it is a pure compound.

To understand the origin of these spectra, which are very different from conventional optical spectra, we need to look at the spin of nuclei (**Figure 18**).

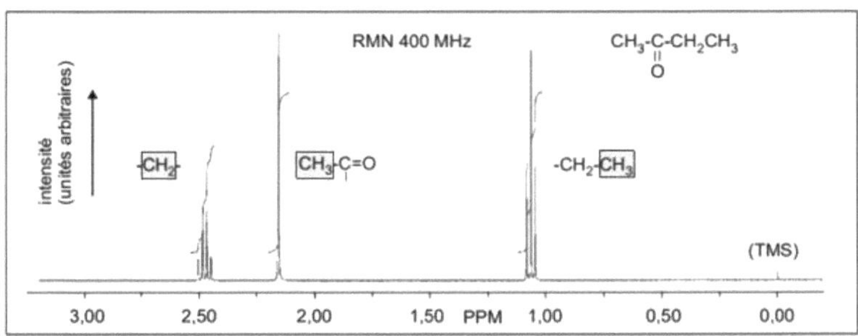

Figure 18 Conventional presentation of an NMR spectrum of hydrogen atoms in an organic compound.

Here we show the spectrum of butanone [CH3 (C = O) CH2CH3] with the integration curve superimposed, allowing the relative areas of the main groups to be evaluated.

signals identified on the spectrum. The nature of the abscissa scale will be explained later.

(Francis Rouessac, Annick Rouessac, 2004, P 278)

3.5.3 Applications

a. Biochemistry: NMR is a powerful technique in metabolic research. It has become a method of choice for uncovering the dynamics and compartmentalisation of metabolic pathways and networks [Fan TWM, Lane AN. 2016]. In addition to this, NMR is also useful for studying intact biological samples such as creur, kidney and skeletal muscle with the P isotope .^{31}

b. Pharmacy: NMR spectroscopy can be used to visualise individual atoms and molecules in various liquids as well as in the solid state [Diehl B. 2008]. It is non-destructive and provides information on molecular structure, enabling structure elucidation and quantification of various organic molecules, as well as providing information on the chemical structure and dynamics of organic molecules in biological systems [Misra G. 2019]. These structural studies provide information related to the functions of organic molecules such as amino acids, proteins, carbohydrates, and antibiotics such as ciprofloxacin, azithromycin and valinomycin.

c. Chemistry: NMR spectroscopy is used unequivocally to identify new compounds, and as such is generally required by scientific journals to confirm the identity of newly synthesised compounds.

d. NMR has been used to study membrane transport systems in vivo or synthetically. For example, it has been used to study the transport of Na^+ ions in human red blood cells.

e. NMR has been used to quantitatively determine the concentration of metabolites. One example is the use of NMR to determine the concentration of phosphocreatine in human muscle [Upadhyay A, et al., 2009].

f. NMR is a very useful technique for identifying and quantifying hydrocarbons in the oil industry. For example, proton nuclear magnetic resonance (H^1) and (C^{13}) are used in the quantitative measurement of liquid hydrocarbons in FACE petrol. It offers good spectral resolution, which is useful for quantifying liquid hydrocarbons in FACE gasoline (Ure AD, et al., 2019).

Chapter 2

Fractionation methods

4.1 Mechanical separation :

4.1.1 Solid-solid separation by dimensions

4.1.1.1 The performance of a classifier depends, according to luckie et ai. (1980) :

a- the feed material (particle size distribution in relation to the cut-off mesh, shape and density, rheological properties),

b- the classifier's characteristics (type of classification, geometric configurations, adjustment options, materials used in its manufacture),

c - operating conditions (ambient temperature and humidity, solids concentration, available energy, feed rate).

These parameters can be correlated (the rheology of the pulp will depend not only on the material but also on the percentage of solids and the shear rate in the classifier).

4.1.1.2 Screening

Screening is a unitary operation for classifying the dimensions of grains of material of various shapes and sizes, by presenting these grains on "perforated" surfaces which allow grains smaller than the dimensions of the perforation to pass through, while larger grains are retained and removed separately. Screening techniques are used in many fields. Screening surfaces can be made from a wide variety of materials, depending on the mesh size required and the materials to be separated: bars, rods, woven metal wires, natural or artificial woven fabrics, expanded foam, etc.

4.1.2 Choice of surfaces The criteria for choosing a screening surface according to a specific service to be provided are :

a- strength (non-deformability, resistance to wear, etc.),

b- regularity of openings,

c- the percentage of the total surface area covered by passageways,

d - low clogging capacity (obstructions due to humidity),

e - poor resistance to obstructions caused by dowelling.

4.1.3 Main types of equipment

The diversity of applications is such that manufacturers have been obliged to create a very wide range of equipment to meet them. The breakdown of the various models according to the granulometries to be cut.

4.1.4 Hydraulic or pneumatic classification

Hydraulic (or pneumatic) size classification refers to all the processes used to separate the solid particles of a suspension in a liquid or gaseous medium into two or more batches of different granularity by the sole action of an acceleration field (gravitational or centrifugal).

4.1.4.1 The main hydraulic classification processes are :

(1) single gravity settling, (2) successive settling, (3) fluidised bed settling, (4) centrifugation (Houot Robert, 1993).

4.1.4.1.1 Decanting

Figure 19. Different types of separating funnel

This is done using separating funnels. There are several types of separating funnel (Fig. 19). Those with the tubing above the tap are the most commonly used, as they give a better view of the interface and therefore enable the two phases to be separated more easily.

4.1.4.1.2 Centrifugation

Centrifugation is a technique that separates compounds in a mixture according to their density under the action of centrifugal force. A precipitate (pellet) and a supernatant are recovered. The mixture to be separated may consist of two liquid phases or solid particles suspended in a liquid. Ultracentrifugation uses even higher rotation speeds (up to 75,000 revolutions per minute) and allows the sedimentation of ultra-microscopic particles.

4.1.4.1.2.1 Principle

Centrifugation is used to separate components of very different size and mass from a liquid (Fig 20). The constituents contained in a sample are subjected to two forces:

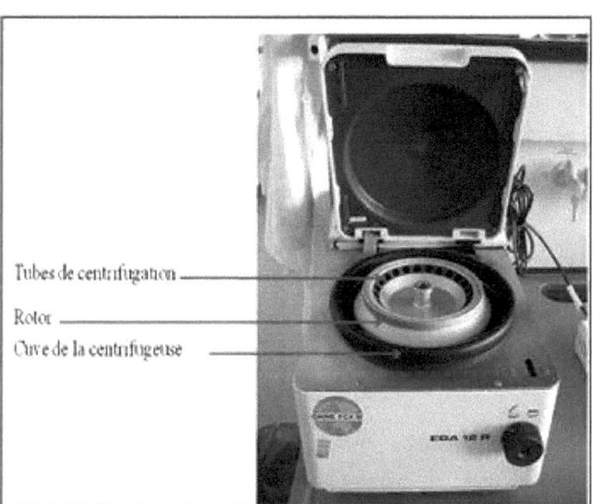

Figure 20. centrifuge.

a-Gravity: This is the force exerted from top to bottom.

b-Archimedes' buoyancy: This is the force exerted from the bottom upwards.

For a given rotational speed, each rotor has a relative centrifugal force in x.g (relative gravitational force or acceleration) which can be expressed in rotational speed in revolutions per minute using the mathematical conversion formula. This formula is:

$\mathbf{g = 1.119 \times 10^{-5} \times r \times N^2}$ where g is the relative centrifugal force, r is the radius of rotation of the rotor (in cm) and N (rotations per minute: rpm) expresses the speed of rotation.

4.2 Separation by diffusion :

4.2.1 Essential oil extraction methods

4.2.1.1 Steam extraction

This is one of the official methods for obtaining EOs (Figure 21) [European Pharmacopoeia (2007)]. In this extraction system, the plant material is subjected to the action of a stream of steam without prior maceration. The vapours, saturated with volatile compounds, are condensed and then decanted into the essencier, before being separated into an aqueous phase (HA) and an organic phase (HE). The absence of direct contact between the water and the plant material, and then between the water and the aromatic molecules, avoids certain hydrolysis or degradation phenomena that could harm the quality of the oil. In addition, the fragrance of the EO obtained is more delicate and distillation, which is regular and faster, means that the top notes are rich in esters [**Raaman, N. (2006)**].

The so-called "top" fractions, highly volatile fragrances due to light molecules, appear first. In most cases, half an hour is enough time to collect 95% of the volatile molecules, which is sufficient for the needs of industry and perfumery, as in the case of lavender. In aromatherapy, however, the process must be prolonged as long as necessary to recover all the volatile aromatic components [**Kaloustian, J., & Hadji-Minaglou, F. (2012)**, **Masango, P. (2005)**, **Gavahian, M., & Chu, Y. H. (2018)**].

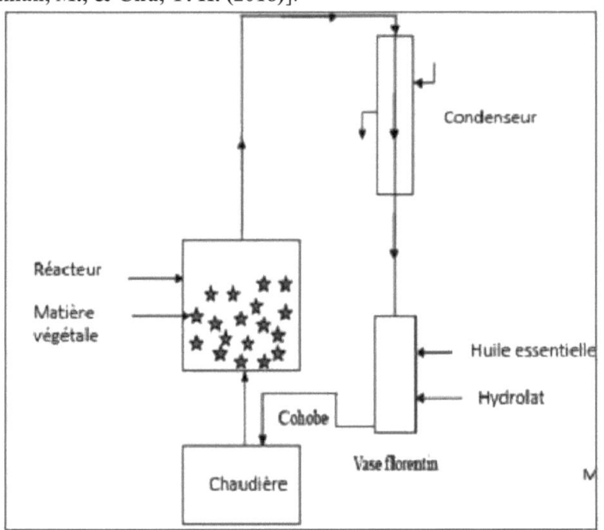

Figure 21. Schematic principle of steam-extraction (EVE) (**Farhat, A. (2010)**)

4.2.1.2 Extraction by Hydrodistillation

It consists of immersing the raw material in a water bath and bringing the whole to the boil (Figure 22, 23). It is generally carried out at atmospheric pressure. Distillation can be carried out with or without cohobing of the aromatic waters obtained during decantation. Some plant organs, particularly flowers, are too fragile to withstand treatment by steam distillation and hydrodistillation (HD) [**Farhat, A. (2010)**].

However, direct contact of EO constituents with water causes chemical reactions leading to changes in the final composition of the extract [**Raaman, N. (2006), Walton, N. N. J., & Brown, D. D. E. (1999)**].

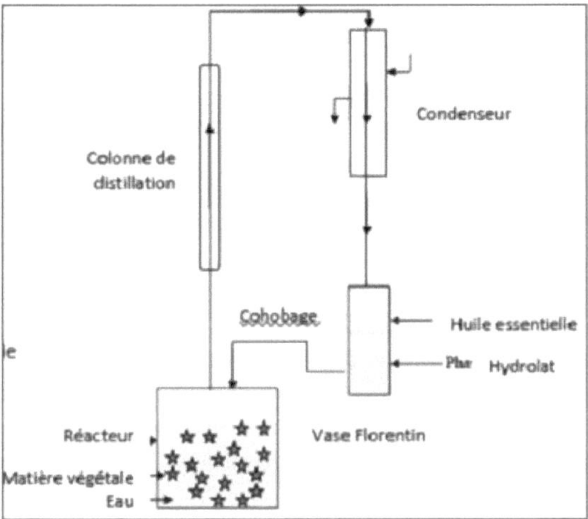

Figure 22. Schematic principle of hydrodistillation (HD) (**Farhat, A. (2010)**)

The operating conditions and, in particular, the duration of distillation have a considerable influence on the yield and composition of the EO. This is why mathematical models are now being developed to optimise these conditions in order to produce EO in a reproducible way.

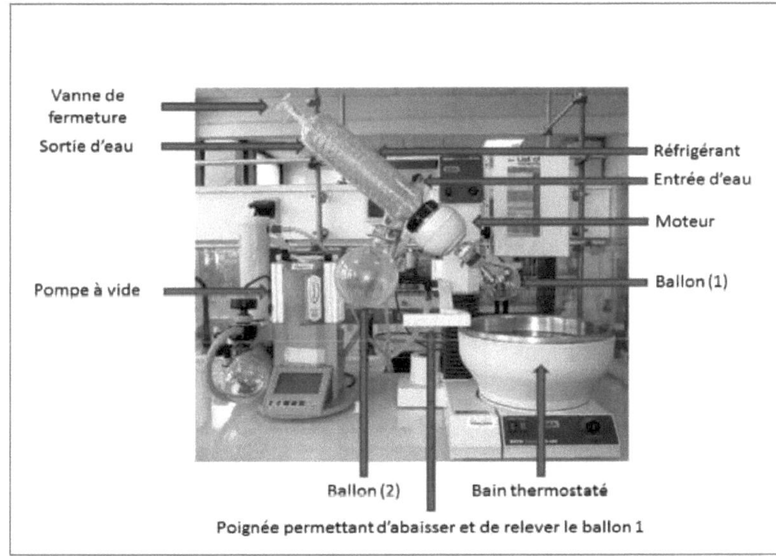

Fig 23. A rotary evaporator (Clément de Mecquenem et al., 2018)

The lability of EO constituents explains why the composition of the product obtained by HD is most often different from that of the mixture initially present in the plant's secretory organs [**Lucchesi, M. E. (2005), Boukhatem M.N. (2018)**].

Hydrodistillation has its limits. Prolonged and powerful heating causes certain plants to deteriorate and certain aromatic molecules to break down. Water, acidity and temperature can induce ester hydrolysis, as well as rearrangements, isomerisations, racemisations and/or oxidations [**Bruneton, J. (1999)**]. The wide variations in composition highlighted by the literature review on EO are therefore easy to understand.

4.2.1.3 Extraction by organic solvent

The solvents most commonly used today are hexane, cyclohexane, ethanol, and less frequently dichloromethane and acetone. As well as being permitted, the chosen solvent must be stable to heat, light and oxygen. Its boiling temperature should preferably be low to facilitate its elimination, and it should not react chemically with the extract. Extraction is carried out using a Soxhlet apparatus or a Lickens-Nickerson apparatus (Figure 24). These solvents have a higher extraction power than water, so that the extracts contain not only volatile compounds, but also a large number of non-volatile compounds such as waxes, pigments, fatty acids and many other substances [**Hubert, R. (1992)**].

Depending on the technique and solvent used, we obtain hydrolysates (water as solvent), alcoholates (diluted ethanol), tinctures (ethanol/water), resinoids (concentrated ethanolic extracts) and concretes (cold and hot extracts using various solvents) [**Hernandez Ochoa, L. R. (2005)**].

The 'classic' solvent extraction technique involves placing a volatile solvent and the plant material to be treated in an extractor. Through successive washings, the solvent is loaded with aromatic molecules, before being sent to the concentrator to be distilled at atmospheric pressure.

The restrictive use of volatile organic solvent extraction is justified by its cost, safety and toxicity issues, and environmental protection regulations. However, yields are generally higher than with distillation, and this technique avoids the hydrolysing action of water vapour [**Lucchesi, M. E. (2005)**].

Faced with this situation, two new techniques have been developed in recent years for distilling flavouring substances from plants: microwave-assisted extraction and supercritical CO_2 extraction.

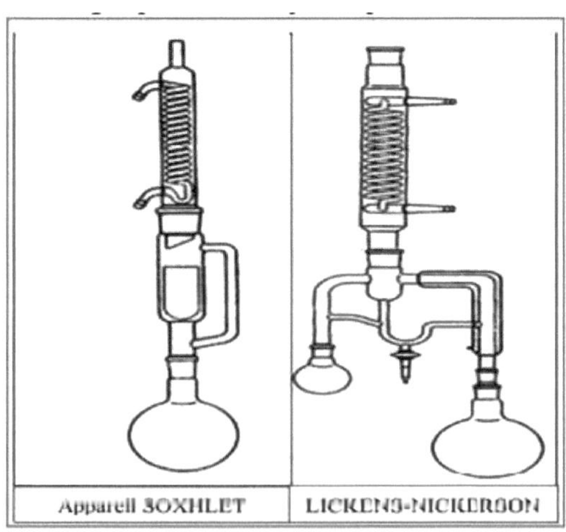

Figure 24: Solvent extraction set-up

4.2.1.4 Microwave-assisted extraction

The advantage of this process is that it considerably reduces distillation time and increases yield. However, no industrial developments have been made to date. Microwave-assisted distillation is currently the subject of a great deal of research and is constantly being improved because it offers many advantages: green technology, energy and time savings, reduced initial investment and minimised thermal and hydrolytic degradation [**Lucchesi, M. et al, (2004), Olivero-Verbel, et al, (2010)**].

The use of microwaves is also a rapidly developing extraction method in its own right. For example, SFME (Solvent Free Microwave Exatrction) is an original combination of microwave heating and dry distillation techniques. It involves placing plant material in a reactor in a microwave oven without adding water or solvent (Figure 25). The internal heating of the water contained in the plant dilates its cells and leads to the rupture of the glands and the oleiferous receptacles. The EO thus released is evaporated with the plant water [**Wang, Z., et al, (2006)**].

Compared with traditional hydrodistillation, SFME is characterised by a reduction in energy consumption and CO_2 emissions but, above all, by an extraction time that is around 9 times faster. The EOs produced by this process contain a higher proportion of oxygenated compounds, with more significant odour values, while monoterpenes are present in smaller quantities [**Ferhat, M. et al, (2006), Golmakani, M. T., & Rezaei, K. (2008)**].

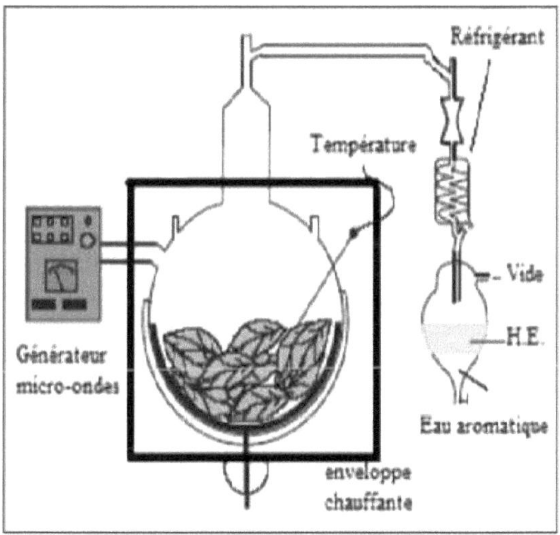

Figure 25: Microwave-assisted hydrodistillation

The experimental protocol for solvent-free microwave-assisted extraction (SFME) is based on three key points:

a-The quantity of plant material was set so as to obtain a sufficient quantity of EO for separation by simple decantation. The aim of this protocol was to avoid the use of organic solvents in order to obtain the cleanest possible product;

b-The microwave power applied (300-450 Watts) during SFME extraction is necessarily a function of the quantity of plant material to be treated. This value represents the amount of power applied in Watts per kilogram of plant material treated;

c-The total extraction time is made up of the heating time (first stage = 10 min) and the extraction time (second stage = 10 min). The heating capacity of microwaves is much greater than traditional heating. The duration of microwave extraction will be considerably reduced compared with conventional hydrodistillation [**Farhat, A. (2010), Ferhat, M. et al., (2008)**].

Here again, preliminary experiments as well as data from the literature [**Chemat, F., et al, (2013).**] have shown that under microwave conditions, unlike a conventional "hydrodistillation" type extraction, it was not necessary to heat for long periods to obtain interesting yields.

Microwaves act on certain molecules, such as water, which absorb the wave and convert its energy into heat. Unlike conventional heating by conduction or convection, the release of heat takes place in the mass. In a plant, for example, microwaves are absorbed by the most water-rich parts of the plant (the vacuoles, the liquid reservoirs of the cells) and then converted into heat. The result is a sudden rise in temperature inside the material, until the internal pressure exceeds the expansion capacity of the cell walls. The steam destroys the structure of the plant cells, and the substances inside the cells can then flow freely outside the biological tissue, and the HE is then carried away by the steam [**Farhat, A. (2010)**].

Lucchesi et al [**2005**] extracted EOs by SFME from three aromatic herbs: basil, mint and thyme. Using this technique, they isolated and concentrated the volatile compounds in a single step, without adding solvent or water. The extracted EOs were richer in oxygenated compounds than the conventional method. In fact, the abundance of oxygenated compounds in the EO is linked to the rapid heating of polar substances with microwaves and the low quantity of water in the medium, which prevents the degradation of compounds by thermal and hydrolytic reactions. This technique offers a number of advantages, such as shorter extraction times, a reduction in the quantity of solvent and very good reproducibility with good yields.

The EO obtained by distillation never exactly represents the aroma and fragrance naturally present in the plant. Microwave-assisted extraction, a new, innovative and environmentally-friendly technique, can solve some of the problems associated with distillation.

4.2.1.5 Supercritical fluid extraction

The originality of the supercritical fluid extraction technique, known as SFE (Fig 26), stems from the use of solvents in their supercritical state, i.e. under conditions of temperature and pressure where the solvent is in a state intermediate to the liquid and gaseous phases and has different physico-chemical properties, in particular increased solvation power. Although many solvents can be used in practice, 90% of EFS are carried out using carbon dioxide (CO_2), mainly for practical reasons. In addition to being easy to obtain due to its relatively low critical pressure and temperature, CO_2 is relatively non-toxic, available in high purity and at low cost, and has the advantage of being easily removed from the extract [**Leszczynska, D. (2007)**].

SFE is a "green" technique that uses little or no organic solvent and has the advantage of being much faster than traditional methods. The chemical compositions of the EOs obtained in this way can differ qualitatively and quantitatively from those obtained by hydrodistillation [**Gomes, P et al, (2007), Peterson, A., et al, (2006), Pereira, C. G., & Meireles, M. A. A. (2010)**].

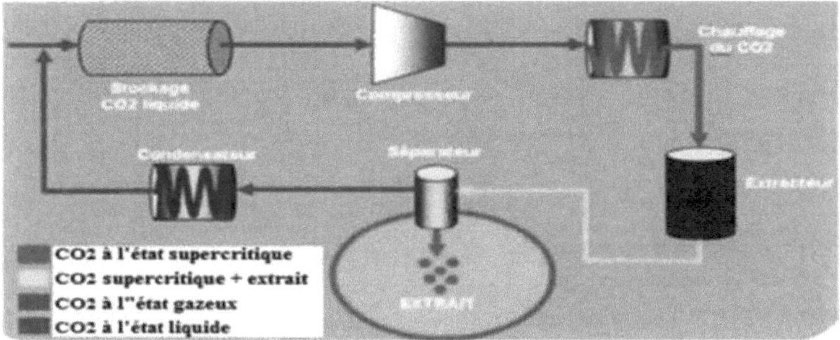

Figure 26 shows the supercritical CO2 extraction technique (as an example).

4.2.2 Chromatography

4.2.2.1 Gas chromatography

This is an analytical chemistry technique that separates volatile or volatilizable compounds without degradation (non-thermolabile). Its separation power exceeds that of all other techniques, at least for essential oils.

Gas chromatography (GC) is a widely used technique (Fig 27). It has a number of advantages, including sensitivity, versatility, rapid development of new analyses and the possibility of automation, all of which increase its appeal (Rouessac. F et al, 2007).

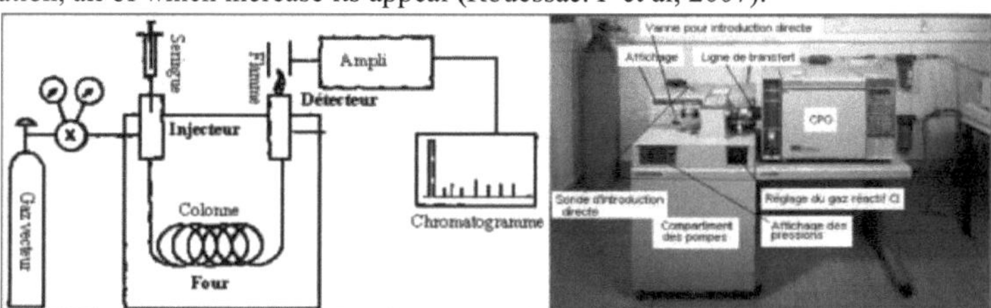

Figure 27: Apparatus for a GPC

The technique has been perfected and can now be used to separate the constituents of very complex mixtures containing up to 200 compounds (Mendham J et al, 2005). It is mainly applied to compounds that are gaseous or can be vaporised by heating without decomposition. It is increasingly used in the main areas of chemistry. The mixture to be analysed is vaporised at the inlet of a column, which contains a solid or liquid active substance known as the stationary phase, and is then transported through the column using a carrier gas. The different molecules in the mixture separate and leave the column one after the other after a certain period of time, which depends on the affinity of the stationary phase for these molecules (Burgot G and Burgot J. L, 2011).

4.2.2.2 Mass spectrometry

It was in 1898 that Joseph John Thomson made the first measurements of the mass-to-charge ratio (*m/z*) of electrons. His theoretical and experimental research into electrical conductivity in gases earned him a Nobel Prize in 1906.

Mass spectrometry is a powerful analytical technique used to quantify known materials, to identify unknown compounds in a sample and to elucidate the structure and chemical properties of different molecules. The process involves transforming the sample into gaseous ions, with or without fragmentation, which are then characterised by their *m/z* ratios and relative abundances **(Fabrice BRAY, 2017)**.

For example, if a sample of methanol (CH3 OH) in the gaseous state is ionised by electron bombardment, a small fraction of the molecules are transformed into charge-bearing species, including the positive ions CH_3**OH**]$^+$. These ions, formed in an excited state, have excess energy, which causes many of them to fragment almost immediately. However, they do not all dissociate in the same way, so that a whole collection of ions with masses lower than that of the original methanol molecules is formed. Generally speaking, these fragments, born of bond breaks and subsequent rearrangement reactions, carry information about the initial molecule (fig. 28 a).

The results are presented in a graph called a mass spectrum, showing the abundances of the ions formed in ascending order of their mass/charge ratio (fig. 28). By operating under identical conditions, fragmentation is reproducible and therefore characteristic of the compound studied. The latter is destroyed by the analysis (Francis Rouessac and Annick Rouessac, 2004).

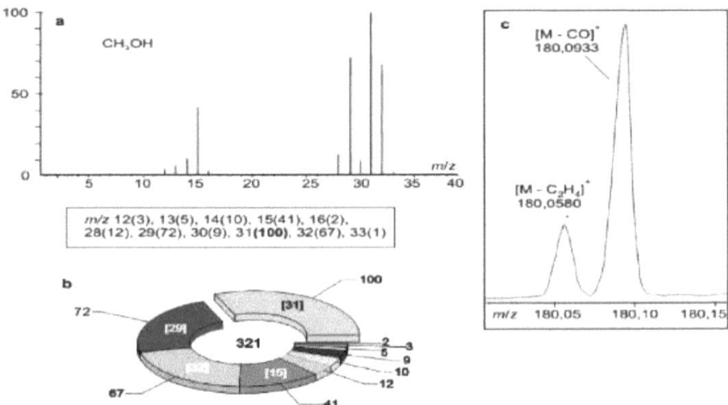

Figure 28 Fragmentation spectrum and mass spectrum presented under graphic or tabular form.

a) Fragmentation spectrum of methanol; b) Unconventional representation of the same spectrum in the form of a circular diagram: statistically, for 321 ions formed, there are 100 of mass 31 u (Da), 72 of mass 29, etc. The various ions constitute as many different populations; c) Part of a high-resolution recording of a compound M showing two ions of similar mass (one by loss of CO and the other of C2H4) (Francis Roux). The various ions constitute as many different populations; c) part of a high-resolution recording of a compound M showing two ions of similar mass (one by loss of CO and the other of C_2H_4) (Francis Rouessac and Annick Rouessac. 2004).

4.2.2.2.1 The mass spectrum can correspond to two very different types of plot:

a-The continuous spectrum (profile) of the selected mass interval. The plot corresponds to a set of signals in the form of peaks of varying widths, depending on the quality of the instrument.

(fig. 28c). These peaks, distributed according to their mass, make it possible to determine the mass of ions with an accuracy of up to 10^{-5} Da for the best instruments. The upper mass limit, which is constantly rising, exceeds 10^6 Da.

b-Fragmentation spectrum ("bar spectrum" or bar chart). This corresponds to the distribution of all the ions formed, grouped together at the nominal masses (Fig. 28) closest to their real masses and presented in the form of vertical lines. The most abundant type of ion leads to the most intense peak, called the base peak, which is given an index of 100. The intensities of the other peaks are expressed as a % of the base peak. This graphic representation of the masses divided into populations, whose heights are proportional to their abundances, corresponds to a histogram (fig. 28a,b). Such diagrams are easy to archive and compare for identification purposes. The disadvantage of this standardised presentation, which leads to the type of plot most commonly used in analysis, is that the same nominal mass may correspond to ions of different atomic composition (Francis Rouessac and Annick Rouessac, 2004).

4.2.2.2.2 The mass spectrometer consists of 3 main parts (Figure 29):

a-The ionisation source, which allows the molecules to pass into the gas phase and be ionised,
b-The analyser, which separates the ions according to their *m/z* ratio,
c-The detector, coupled to a computer system, which processes the data,

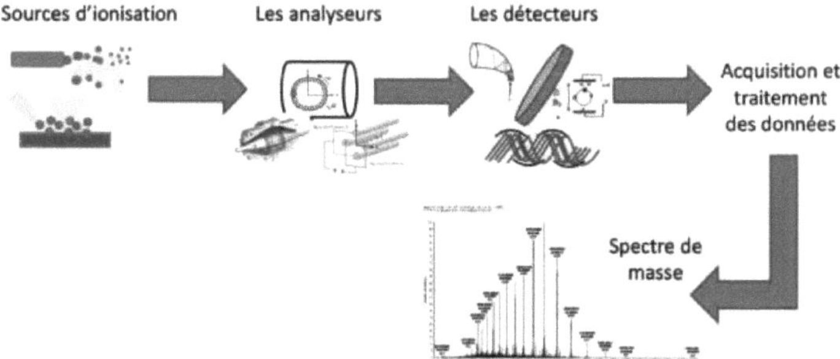

Figure 29: Schematic representation of a mass spectrometer (Fabrice BRAY, 2017).

There is a great diversity of mass spectrometry equipment, due to the wide variety of sources, analysers, types of coupling and fragmentation. Of course, not all configurations are feasible. These variable instrument configurations allow samples of different natures to be analysed and a variety of information to be obtained. Resolution and sensitivity will differ from one instrument to another because of their configuration.

The two essential instrumental characteristics for measuring ions by mass spectrometry are resolving power and mass measurement accuracy. Good spectral resolution makes it possible to determine the charge state of ions, access their isotopic distribution and discriminate between isobaric ions. This resolution is also linked to the accuracy of the mass measurement, which corresponds to the sharpness of the peak observed (**Fabrice BRAY, 2017).**

4.2.2.3 Analysis by coupling gas chromatography to mass spectrometry (GC/MS)

Gas chromatography-mass spectrometry (Fig 30) is an analytical method that combines the performance of gas chromatography and mass spectrometry to accurately identify and/or quantify many substances. The method is based on the separation of constituents using GC and their identification using MS.

Gas chromatography separates the molecular fractions making up the sample on the basis of the speed of movement and retention time taken to travel through a column filled with a stationary phase. Mass spectroscopy uses energy sources to ionise, fragment and finally separate molecular groups according to the mass/electric charge ratio (m/q) (Naira PEREZ VASQUEZ, 2015).

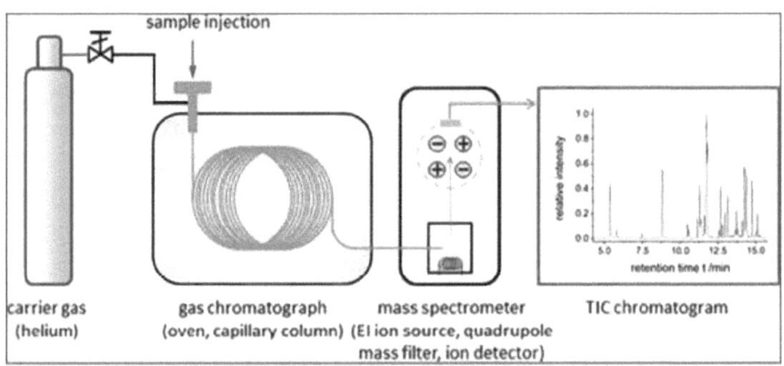

Figure 30: GC/MS coupling diagram

The combination of these two GC/MS analysis techniques enables the components of the sample to be separated and each component to be identified, providing a complete qualitative and quantitative analysis of the product to be analysed. Identification is then carried out by comparing the retention indices (Ir) and spectral data (mass spectra) of the individual components with the characteristics of reference products contained in spectral libraries. The advantage of chaining a chromatographic interface to a spectrometer is the ability to analyse the individual spectrum of a compound.

It is the most widely used technique for analysing essential oils, largely due to the ease of use of high-performance separation and detection systems, and relatively low cost.

The mass spectrometer must perform the following operations: 1- vaporise; 2- ionise; 3- measure the m/z ratios.

4.2.2.3.1 Classically, the mass spectrometer is made up of 3 main parts:

a-The ionisation source, whose role is to bring the molecules into the gas phase and ionise them.

b- The analyser, which separates ions according to their m/z ratio.

c- The detector, which is coupled to a computer system to process the data and produce mass spectra (Naira PEREZ VASQUEZ, 2015).

4.2.2.3.2 Main modes of analysis - examples of applications :

The combination of gas chromatography and mass spectrometry (GC-MS) is now at its peak, with applications in fields as varied as the food industry, medicine, pharmacology and the environment (Tranchant J., 1995). The GC-MS technique is reserved for the analysis of molecules that are easily vaporised and thermally stable, i.e., to a first approximation, compounds of low to medium molecular weight (less than 700 Da). Given these limitations, GC-MS is a formidable analytical tool. The diversity of injection methods and capillary columns (geometry, nature of the stationary phase) means that extremely complex mixtures (essential oils, metabolites, hydrocarbons, etc.) can be separated (Stéphane Bouchonnet and Danielle Libong, 2014). Applications include the detection of chemical weapons [SYAGE (J.A.), et al, 2001] [HOOIJSCHUUR (E.W.), et al, 2002], explosives [FIALKOV (A.B.) and AMIRAV (A.), 2003] for environmental control [MARRIOTT (P.J.), et al, 2003] [SANTOS (F.J.) and GALCERAN (M.T.), 2003] or space research [NIEMANN (H.B.), et al, 2002].

7-2-4 High-performance liquid chromatography (HPLC)

HPLC analysis allows compounds in solution to be separated by eluting them through a chromatographic column using a liquid mobile phase, which is itself percolated at high pressure. There is a real three-way interaction between the analyte, the stationary phase and the mobile phase, based on the physico-chemical affinity between the three **(BURGOT and BURGOT, 2006; MARCOZ, 2003).**

This method can be used to separate compounds of varying molar mass and different chemical nature. HPLC has become the leading analytical technique used in research and analysis laboratories. It covers all fields of application **(SAUNIER and GODIN, 2013).**

4.2.2.3.3 Principle

The mobile phase passes through a tube called a column, which may contain porous granules (packed column) or be covered on the inside by a thin film (capillary column). In both cases, the column is called the stationary phase.

The mixture to be separated is injected at the inlet of the column where it is diluted in the mobile phase, which carries it through the column. This mixture must be pushed at high pressure to ensure a constant flow rate through the column and avoid any pressure drop. **(CUQ, 2007).**

If the stationary phase has been chosen correctly, the constituents of the mixture, generally known as solutes, are retained unevenly as they pass through the column.

The result of this phenomenon, known as retention, is that the constituents of the injected mixture all move more slowly than the mobile phase and their speed of movement is different. They are thus eluted from the column one after the other and separated. Each solute is then subject to a retention force, exerted by the stationary phase, and a mobility force, due to the mobile phase. **(PANAIVA, 2006).**

A detector placed at the exit of the column, coupled to a recorder, makes it possible to obtain a plot called a chromatogram. It sends a constant signal called the baseline to a recorder in the presence of the carrier fluid alone; as each separate solute passes through, a peak is recorded over time.

Under given chromatographic conditions, the "retention time" (the time after which a compound is eluted from the column and detected) characterises a substance qualitatively. The amplitude of these peaks, or the area bounded by these peaks and the extension of the baseline, is used to measure the concentration of each solute in the injected mixture.

The principle is therefore to exploit the interactions between the solutes and the two phases (mobile and stationary) to separate these solutes according to their affinities and thus identify and/or measure them **(PANAIVA, 2006).**

Conventional high-performance liquid chromatography involves solute, mobile phase and stationary phase exchange mechanisms, based on partition or adsorption coefficients depending on the nature of the phases present **(MARCOZ, 2003).**

4.2.2.3.4 Instrumentation

According to the following authors: **Saunier and Godin, (2013),** in all high performance liquid chromatography equipment, there is a set of modules linked together by small diameter tubes illustrated in the diagram below (Fig 31).

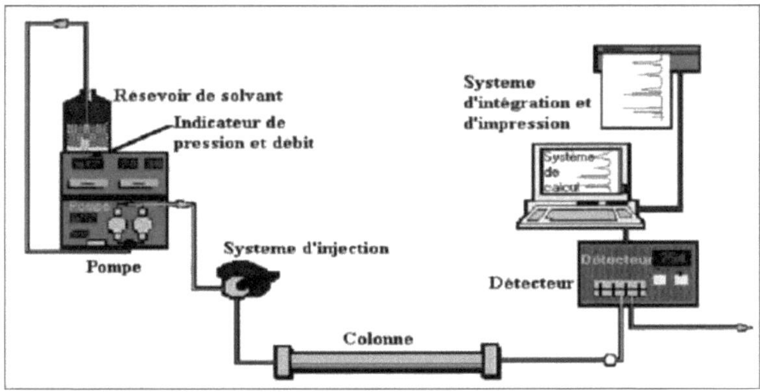

Diagram 31: HPLC instrumentation.

4.2.2.3.4.1 Solvent tanks

They contain the mobile phase in sufficient quantity, and several vials of eluents (solvents of different polarities) are available so that elution gradients (mixtures of several solvents at variable concentrations) can be produced using the pumps **(SALGHI. R, 2004).**

4.2.2.3.4.2 Pumps

The pumps used in HPLC deliver solvents at a constant flow rate under high pressures of up to a few hundred bar, the aim of which is to force the mobile phase through the column. Their role is twofold:

Firstly, they enable a constant flow rate to be maintained, whatever the pressure and the variation in composition of the mobile phase, but also to obtain a sufficient flow rate of up to 10 ml/min without any loss of accuracy and reproducibility. **(SAUNIER and GODIN, 2013)**

The pumps are fitted with a gradient system that allows the nature of the solvent to be programmed. They therefore make it possible to work in isocratic mode (with 100% of the same eluent throughout the analysis) or gradient mode (a variation in the concentration of the constituents of the eluent mixture). Current pumps have variable flow rates ranging from a few µť to several ml/min **(THIERRY, B .2009).**

Note: Gradient elution is more expensive than isocratic elution and is preferred for single samples containing less than 10 components **(SCHELLINGER ET AL, 2006).**

4.2.2.3.4.3 Injection valve

This is an injector with sampling loops. There are loops of different volumes: 10, 20, 50 µť. The loop volume is chosen according to the size of the column and the assumed concentration of the products to be analysed. The injection loop system makes it possible to have a constant injected volume, without significant variation in pressure, in the circuit from the pumps to the column inlet; this is important for quantitative analysis. **(THIERRY. B, 2009); (LEBIHAN and LEMASSON, 2008)**

4.2.2.3.4.4 Columns

The tube must be made of stainless steel or glass (inert to chemicals). It has a constant cross-section, with a diameter of between 4 and 20 mm and lengths generally between 15 and 30 cm. It contains the stationary phase, which may be polar (normal phase) or non-polar (reverse phase). **(THIERRY, B .2009) ; (LEBIHAN et LEMASSON, 2008)**

The column supports are very finely divided solids whose main property is that they are inert with respect to the stationary and mobile phases. The most commonly used support is silica gel, made up of grains with diameters ranging from 1.5 to several tens of micrometres. These grains have pores of different diameters (80 to 300 μm) on their surface through which the mobile phase flows. **(SAUNIER and GODIN, 2013)**

a-The normal phase: this is made up of silica gel; this material is very polar, so an apolar eluent must be used. When a solution is injected, the polar products are retained in the column, unlike the apolar products, which come out at the top **(PREMA, 2003).**

The disadvantage of such a phase is the rapid deterioration of the silica gel over time, which leads to a lack of reproducibility in separations. **(THIERRY, B .2009)**

b-The reverse phase: this is mainly composed of silica grafted with linear chains of 8 or 18 carbon atoms (C8 and C18). This phase is apolar and therefore requires a polar eluent (acetonitrile, methanol, water). In this case, the polar compounds are eluted first.

Unlike a normal phase, there is no change in the stationary phase over time and the quality of the separation is therefore maintained constant **(JACOB. V, 2010); (FEKETE et al, 2009); (THIERRY, B .2009).**

The reverse phase HPLC technique is often chosen as the initial choice. It is increasingly considered to be the best separation technique for achieving high resolution, short times and better reproducibility of retention times by manipulating HPLC conditions **(WANG. *et al*, 2003); (MAJORS. *et al*, 2002).**

In general, there are many factors that influence column separation and efficiency, including temperature, particle size, mobile phase length and flow rate **(WANG *et al*, 2003); (SNYDER. *et al*, 1988).**

4.2.2.3.4.5 Detectors

Placed at the exit of the column, the detector records a signal (electrical or optical) as a function of time, characteristic of the progressive passage of solutes.

Two types of detectors are typically used:

a-The UV-visible detector: this measures the absorption of light by the product as it leaves the column and operates at a constant wavelength, which is set by the operator. The deuterium lamp is used for wavelengths varying from 190 to 350 nm and the mercury vapour lamp is used at the non-variable wavelength of 254 nm **(JACOB. V, 2010); (THIERRY, B .2009); (LEBIHAN and LEMASSON, 2008)**.

According to **Saunier and Godin, (2013),** for this type of detector to be usable, it is necessary that:

b-The product to be detected absorbs light at a wavelength accessible to the device and if its absorption coefficient is sufficiently large, the mobile phase does not absorb light at the wavelength chosen by the operator.

4.2.2.3.4.6 The refractometer :

It measures the variation in the refractive index of the liquid as it leaves the column. This measurement is extremely accurate, but depends on the temperature of the liquid. This index is compared with that of the pure mobile phase: there is therefore a reference, hence the term index variation.

This detector excludes variations in the composition of the mobile phase; it is therefore only possible to work in isocratic mode with this detector. Data is collected using either an integrator or an acquisition station (Fig 32) **(JACOB. V, 2010); (THIERRY, B .2009); (LEBIHAN and LEMASSON, 2008).**

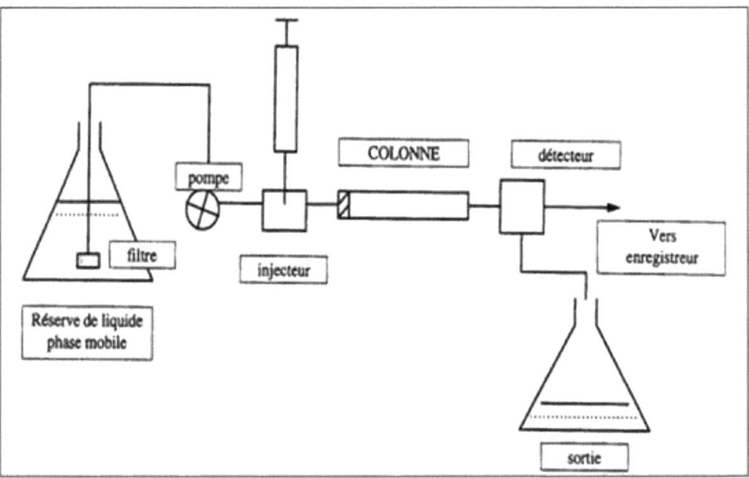

Figure 32: HPLC operating principle

4.2.2.3.5 Chromatogram analysis

A good separation will result in a distinct separation of the peaks corresponding to each of the products. Figure 33 shows the difference between a good and a bad chromatogram.

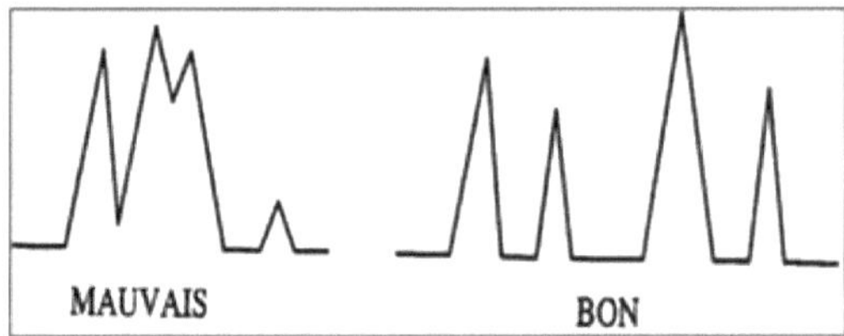

Figure 33: Qualitative analysis of a chromatogram.

A chromatogram must be perfectly reproducible, so for each analysis you need to specify the type of column: brand, type, diameter, length, support, etc., the type of eluent: solvent, if it is a mixture specify its composition, its flow rate, the detection mode λ in nm, the quantity injected, the start of the injection on the chromatogram and the sensitivity of the detector.

4.2.2.3.6 Qualitative analysis

a-Retention time

The retention time (t_R in min) is a characteristic of each solute under the set operating conditions. The greater or lesser interaction between the mobile phase and the stationary phase (normal or reverse polarity) has an effect on the retention times of the solutes.

Retention time (t_R) = time elapsed between injection and the maximum point of the peak.

Dead time (t_M) = time elapsed for a non-retained component to pass through the column.

The retention time is usually used instead of the retention volume and its size depends on the nature of the stationary phase, the nature of the mobile phase, the flow rate of the mobile phase and the length of the column.

4.2.2.3.7 Quantitative analysis

Chromatography is a quantitative analytical technique, based on a relationship between the mass quantity of the solute injected and the area of the chromatogram peak given by the detector, i.e. the area of the chromatographic peaks is proportional to the concentration or quantity of the product analysed.

The mass m_i of solute *i* detected is proportional to the area *Ai* of the signal measured by the detector:

$mi = Ki.Ai$

Ki is the response coefficient of the detector for solute *i*. It depends on the column used, the sensitivity of the detector to solute *i* and the experimental conditions.

Ai is the area of the elution peak of solute i on the chromatogram. This peak area is determined automatically using a recorder-integrator.

In practice, *Ki* and *Ai* must therefore be determined for a given solute *i* under the given analysis conditions.

4.2.2.3.8 Advantages of high-performance liquid chromatography:

a-It is used for solutes with a high molecular weight, low volatility and sensitivity to high temperatures. These solutes may be non-polar, polar or ionic.

b-It analyses very small quantities, is extremely sensitive, has high separation power and excellent reproducibility (JACOB. V, 2010; THIERRY, B .2009).

4.2.2.3.9 HPLC applications

HPLC has an extremely wide range of applications, but it is particularly used in cosmetology and biochemistry. It can be used to analyse thermally unstable substances, since the operation is carried out at room temperature, or low-volatility substances with a molecular weight of up to 2,000,000, or ionised substances. It is therefore a 'soft' method for molecules. Molecules of biological interest, such as vitamins, sugars and acids, can be analysed directly without the formation of derivatives, and proteins and synthetic polymers can be separated, even if their mass is high. Some examples of applications:

a-Kinetics of chemical reactions (OSTROWSK I. T *ET AL*, 2003) or enzymatic reactions. (GHASHGHAIE .J ET AL, 2001) ; (CHEVIRON .N, 2000)

b-Calculation of the partition coefficient for series of analogues (GUENARD .D. *Et Al*, 2000)

c-Study of the phospholipophilic characteristics of a family of organic molecules by HPLC .

d-Dosing of active molecules in biological fluids or crude extracts (ADELINE. MT *ET AL*, 1997)

e-Analytical study of biologically active extracts of natural substances (POUPAT. C. *Et Al*, 2000)

f-Isolation of pure products from mixtures of natural or synthetic products (DUMONTET. V *ET AL*, 2004)

HPLC is a physico-chemical method for detecting and quantifying residues of a fairly wide range of antibiotics, covering all the families used in human and veterinary medicine. It is a much more selective and sensitive method than microbiological methods, as it allows the molecules to be identified separately and therefore avoids the problems of possible interference between substances. (BOATTO *et al*, 1998)

Various extraction methods can be used, including liquid-liquid extraction, ultrafiltration, protein precipitation and solid-phase extraction (OLIVEIRA *et al*, 2006).

Chapter 3

Marking methods

5.1 Enzyme-linked immunosorbent assay (**ELISA**)

Enzyme-linked immunosorbent assay (ELISA) is frequently used to measure the presence and/or concentration of an antigen, antibody, peptide, protein, hormone or other biomolecule in a biological sample. It is extremely sensitive, capable of detecting low concentrations of antigen. The sensitivity of ELISA is attributed to its ability to detect interactions between a single antigen-antibody complex (Porstmann, T. and Kiessig S.T. 1992). In addition, the inclusion of a zymatically conjugated antigen-specific antibody allows the conversion of a colourless substrate into a chromogenic or fluorescent product that can be detected and easily quantified by a plate reader. From the values generated by titrated amounts of a known antigen of interest, the concentration of the same antigen in experimental samples can be determined. Different ELISA protocols have been adapted to measure antigen concentrations in a variety of experimental samples, but they all have the same basic concept (Suleyman Aydin. 2015). The choice of which type of ELISA to perform, indirect, sandwich, or competitive, depends on a number of factors, including the complexity of the samples to be tested and the available antigen-specific antibodies to be used. The indirect ELISA is frequently used to determine the results of an immunological response, such as measuring the concentration of an antibody in a sample. The sandwich ELISA is best suited to analysing complex samples, such as tissue culture supernatants or tissue lysates, where the analyte, or antigen of interest, is part of a mixed sample. Finally, the competitive ELISA is most often used when there is only one antibody available to detect the antigen of interest. Competitive ELISAs are also useful for detecting a small antigen with only a single antibody epitope that cannot accommodate two different antibodies due to steric hindrance.

5.1.1 Type of ELISA methods (OIV-Oeno 427-2010 Modified by OIV-COMEX 502-2012)

5.1.1.1 General protocol for the direct and indirect ELISA method

The direct, one-step method uses only a conjugated antibody which is incubated with the antigen contained in the sample/reference and bound to the surface.

The two-step indirect method uses a conjugated secondary antibody for detection. First, the primary antibody is incubated with the antigen contained in the sample/reference and bound to the well. This is followed by incubation with the conjugated secondary antibody which recognises the primary antibody (**Figure 34**).

a-Direct

1-Prepare a surface to which the sample antigen is bound.

2-Block all non-specific binding sites on the surface.

3-Apply enzyme-linked antibodies that bind specifically to the antigen.

4-Rinse the plate to remove excess antibodies (not bound to the antigen).

5-Add a chemical substance which will be converted by the enzyme into colour, fluorescence or an electrochemical signal.

6-Measure the absorbance, fluorescence or electrochemical signal (current) of the plate wells to determine the presence and quantity of antigen.

Before analysis, the antibody preparations must be purified and conjugated.

b-Indirect

1-Prepare a surface to which the sample antigen is bound
2-Block all non-specific binding sites on the surface.
3-Apply enzyme-linked primary antibodies that bind specifically to the antigen.
4-Rinse the plate to remove excess primary antibodies (not bound to the antigen).
5-Apply enzyme-linked secondary antibodies which are specific to the primary antibodies.
6-Rinse the plate to remove excess enzyme-conjugated antibodies (unbound).
7-Add a chemical substance which will be converted by the enzyme into colour, fluorescence or an electrochemical signal.
8-Measure the absorbance, fluorescence or electrochemical signal (current) of the plate wells to determine the presence and quantity of antigen.
Before analysis, the two antibody preparations must be purified and one conjugated.

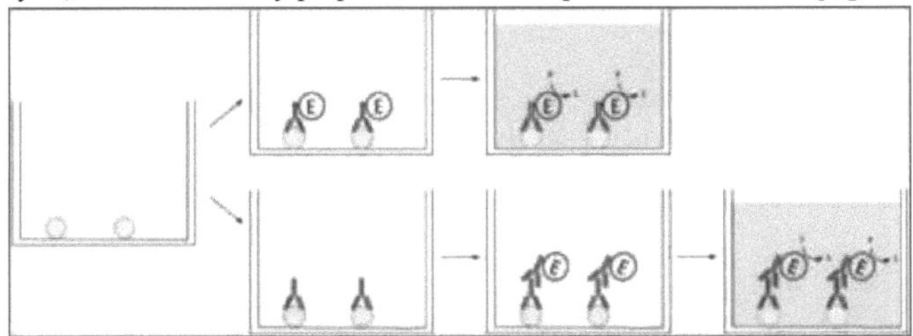

Figure 34: Direct and indirect ELISA

For most applications, a polystyrene microtitre plate with high binding capacity is perfectly suitable; however, consult the manufacturer's instructions to determine the most appropriate type of plate for binding the given antigen.

The main advantage of the direct and indirect ELISA methods is their high sensitivity, achieved by a relatively simple procedure with reduced chances of non-specific binding. However, it is only applicable in cases containing low levels of non-antigenic protein.

5.1.1.2 General protocol for the competitive ELISA method

The term competitive describes tests where the measurement involves quantifying a substance according to its ability to interfere with an established system. Detection can be direct, using the one-step method, or indirect, using the two-step method (SCIPPO, 2006) (**Figure 35**).

a-Direct

1 Prepare a surface to which a known quantity of the desired antigen is bound
2. Block all non-specific binding sites on the surface
3. Apply sample (antigen) or reference and enzyme-linked antibodies
that bind specifically to the antigen on a coated microplate. Antigens immobilised on the surface and antigens in solution compete for antibodies. So the more antigen in the sample, the less the antibody will be able to bind to the immobilised antigen.
4. Rinse the plate to remove excess (unbound) antibodies and complexes
unbound antibody-antigen
5. Add a chemical substance that will be converted into colour by the enzyme,
fluorescence or electrochemical signal
6. Measure the absorbance, fluorescence or electrochemical signal (current) of the
wells of the plate, to determine the presence and quantity of antigen
Before analysis, the antibody preparations must be purified and conjugated.

b-Indirect

1. Prepare a surface to which a known quantity of antigen is bound
2. Block all non-specific binding sites on the surface
3. Apply sample (antigen) or reference and enzyme-linked antibodies
that bind specifically to the antigen on a coated microplate. Antigens immobilised on the surface and antigens in solution compete for antibodies. So the more antigen there is in the sample, the less the antibody will be able to bind to the immobilised antigen.
4. Rinse the plate to remove excess (unbound) antibodies and complexes
unbound antibody-antigen
5. Add a secondary antibody, specific to the primary antibody, conjugated with a
enzyme
6. Rinse the plate to remove excess (unbound) conjugated antibodies
7. Add a chemical substance which will be converted into colour by the enzyme,
fluorescence or electrochemical signal
8. Measure the absorbance, fluorescence or electrochemical signal (current) of the
wells of the plate, to determine the presence and quantity of antigen
Before analysis, the two antibody preparations must be purified and one of them conjugated.

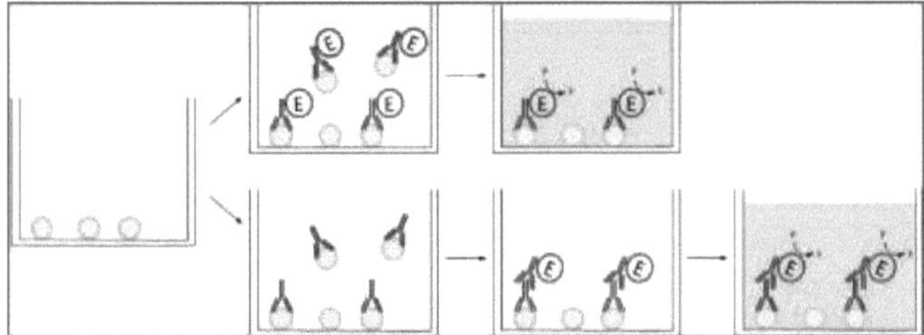

Figure 35: Competitive direct and indirect ELISA

For competitive ELISA, the higher the initial concentration of antigen, the weaker the possible signal.

For most applications, a microplate with high binding capacity is best, however, consult the manufacturer's instructions to determine which type of plate is most suitable for binding the antigen under consideration.

5.1.1.3 General protocol for the Sandwich ELISA method

The sandwich ELISA measures the amount of antigen between two layers of antibodies (i.e. the capture and detection antibodies). The antigen to be measured must contain at least two different antigenic sites (epitopes) to bind two different antibodies. Both monoclonal and polyclonal antibodies can be used (SCIPPO, 2006) **(Figure 36)**.

a-Direct

1. Prepare a surface to which the capture antibody is bound
2. Block all non-specific binding sites on the surface
3. Apply the standard or the sample containing the antigen
4. Wash the plate to remove molecules not recognised by the capture antibody
5. Add enzyme-linked antibodies (detection antibodies) which bind to the enzyme.
specifically to the antigen
6. Rinse the plate to remove excess enzyme-bound antibodies (unbound)
7. Add a chemical substance which will be converted into colour by the enzyme,
fluorescence or electrochemical signal

8. Measuring the absorbance, fluorescence or electrochemical signal (current) of wells of the plate, to determine the presence and quantity of antigen

Before analysis, the two antibody preparations must be purified and one of them conjugated.

b-Indirect

1. Prepare a surface to which the capture antibody is bound
2. Block all non-specific binding sites on the surface
3. Apply the sample containing the antigen
4. Wash the plate to remove molecules not recognised by the capture antibody
5. Add primary antibodies that bind specifically to the antigen
6. Rinse the plate to remove excess (unbound) primary antibodies
7. Add enzyme-linked antibodies (secondary antibodies) which bind to the enzyme. specifically to the primary antibody.
8. Rinse the plate to remove excess enzyme-bound antibodies (unbound)
9. Add a chemical substance which will be converted into colour by the enzyme, fluorescence or electrochemical signal
10. Measure the absorbance, fluorescence or electrochemical signal (current) of the plate wells to determine the presence and quantity of antigen.

Before analysis, the two antibody preparations must be purified and one of them conjugated.

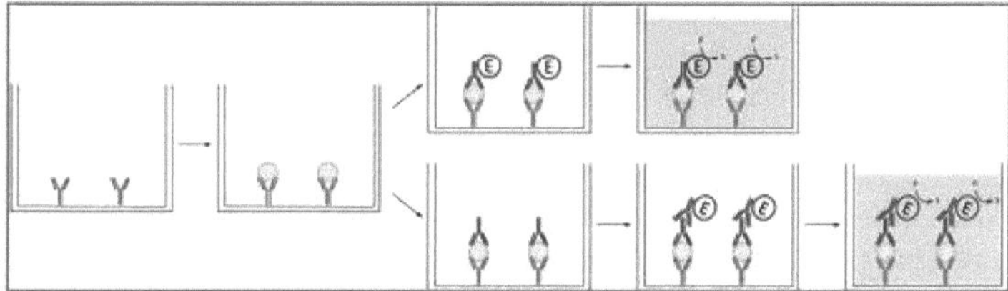

Figure 36: Direct and indirect Sandwich ELISA

For the indirect Sandwich ELISA method, it is necessary for the capture and detection antibodies to be cultured in different species (e.g. mouse and rabbit) so that secondary antibodies attached to enzymes specific to antibody detection do not also bind to the capture antibodies.

For most applications, a microplate with high binding capacity is best, however, consult the manufacturer's instructions to determine which type of plate is most suitable for binding the antigen under consideration.

For the Sandwich ELISA method, the measurement is proportional to the quantity of antigen contained in the samples.

The advantage of the Sandwich ELISA method is that the original samples do not need to be purified before analysis, and the method can be very sensitive.

5.2 Precipitation in a gel medium

5.2.1 Interest :

Precipitation methods in a gel medium are applied to the qualitative analysis of a mixture of Ag in a solution or to the immunochemical assay of an Ag.

5.2.1.1 Double immunodiffusion or Ouchterlony reaction

The Ouchterlony immunodiffusion assay, developed by the Swedish physician Örjan Ouchterlony, is used for the detection of antigens and antibodies, as well as for the determination of homologies between antigens (Ouchterlony O. 1949, Ouchterlony O. 1962).

This assay is also commonly used as a laboratory exercise in undergraduate immunology and microbiology courses to illustrate antigen-antibody precipitation reactions to students (Kindt T, et al., 2007, Armstrong B. 2008). In the Ouchterlony assay, a series of samples (the antigens) are placed in the outer wells of a gel plate, and the antibodies (sera) are placed in the central well, after which they diffuse and form different geometric precipitation lines in the gel (Fig. 37). Intersecting lines can produce full identity (without spurs), partial identity (with one spur), or non-identity when the two lines intersect completely (two spurs), as shown in Fig. 37 (Hornbeck P. 2017, Bailey GS. 1996).

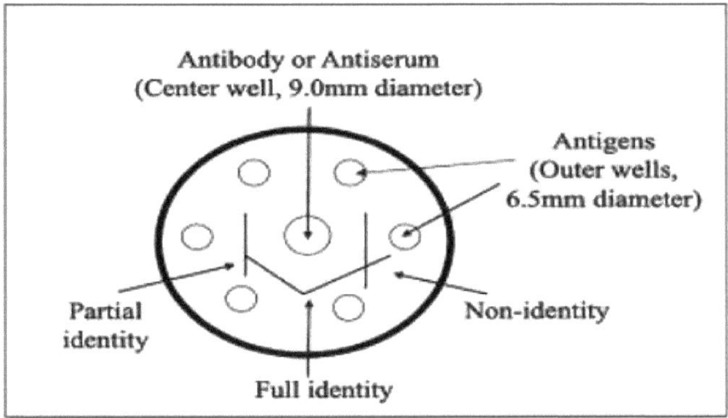

FIG 37. The Ouchterlony immunodiffusion assay and the different geometric precipitation patterns formed in the agarose gel. The serum is placed in the central well and the antigens are placed in the outer wells. Precipitation lines form between the central and outer wells, and depending on the homology between adjacent antigens, different geometric projection patterns form at the intersection, such as full identity (no projection), non-identity (two projections) and partial identity (one projection).

Ag and Ac solutions are deposited in spaced wells in an agarose gel. The molecules diffuse into the gel according to their size, forming precipitation lines for each Ag and Ac system.

Each precipitation line corresponds to the respective equivalence zone, i.e. the formation of an Ag-Ac network.

This method can be used to analyse a mixture of Ag and identify its constituents. When two proteins diffuse in a gel when they meet the Ac, a distinction is made between identity, non-identity or partial identity reactions.

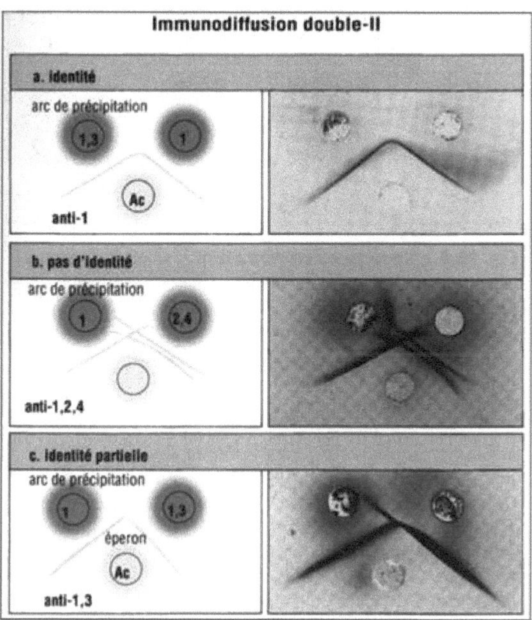

Figure 38: Immunodiffusion on plate (Ouchterlony technique)

The bands merge, but beyond the melting point of the two precipitates, there is a projection that extends the precipitate formed by the antigen. This is the partial identity reaction (Figure 38).

5.2.1.2 Simple radial immunodiffusion procedure (Mancini technique)

2% agarose (SRL, Bombay, India) was prepared in phosphate-saline buffer (PBS), pH 7.4, and transferred to a Petri dish. The dish was cooled to 60°C. Polyclonal serum containing antibodies to M. tuberculosis was then added to the Petri dish (at a concentration of 600 μg of serum protein/3 ml/plate). The agarose with serum was allowed to solidify. Wells were cut out and test samples (cerebrospinal fluid suspected of TBM) at different dilutions were loaded using a micropipette. The gel plates were incubated in a humid chamber for 24 to 48 hours. The diameter of the precipitation rings was measured when the precipitation ring was small, the gel was stained with commossie GR 250 amido black or blue brilliance (0.25 g/100 ml) for 30 minutes and then decolorised with a decolorising solution containing 5:1:5 methanol, acetic acid and water, as reported elsewhere [Mancini G, et al., 1965] (Figure 39, 40).

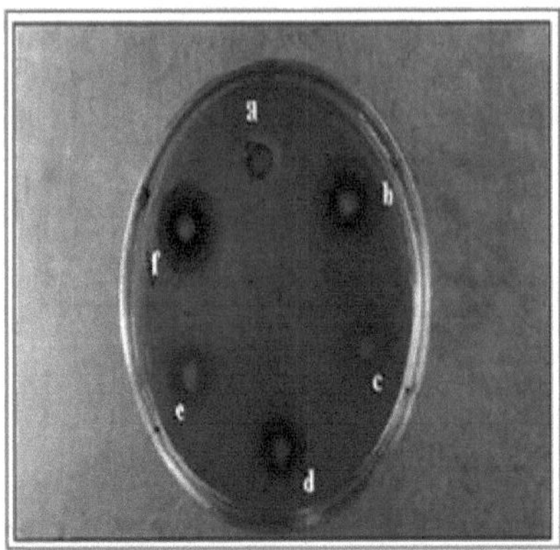

Figure 39. Single radial immunodiffusion (SRID) assay using plates inoculated with polyclonal serum directed against tuberculous meningitis (TBM). Wells b, d and f contained cerebrospinal fluid (CSF) from suspected TBM cases and were negative for acid-fast bacilli (AFB). Wells a, c and e were negative controls from patients suffering from

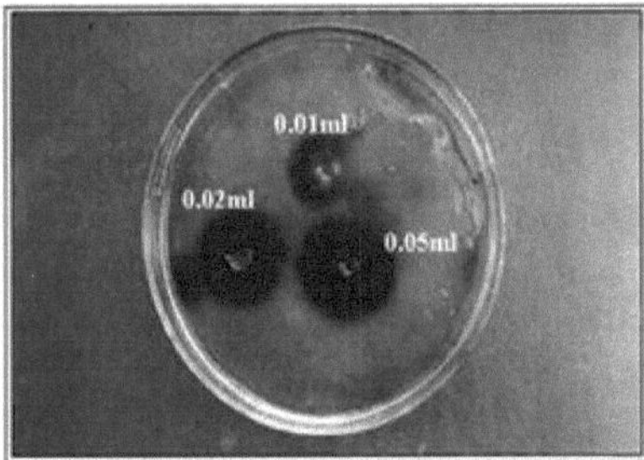

Figure 40. sera produced using CSF with a positive acid-fast bacilli (AFB) test for TBM seeded in the agarose plate. The cell-free antigen for TBM was used in different wells to show the specificity of the serum against TBM bacilli (Rajpal S. Kashyap, et al., 2002).

5.2.1.3 Radial immunodiffusion (Mancini technique)

This method involves incorporating a specific antiserum into the agar and depositing the Ag solution in wells. At equilibrium, a precipitation ring is formed, the square of whose diameter is proportional to the Ag concentration. The concentration is expressed by reference to a standard curve with an Ag of known concentration (figure 41).

5.3 Immunoblot or Western blot

The transfer of proteins or nucleic acids to microporous membranes is called 'blotting', and this term encompasses both 'spotting' (manual deposition of samples) and transfer from planar gels. Proteins that are separated on sodium dodecyl sulphate polyacrylamide gel electrophoresis (SDS-PAGE) gels are usually transferred to adsorbent membrane supports under the influence of an electric current in a procedure known as Western blot (WB) immunoprecipitation or protein transfer (Towbin H, et al., 1979, LeGendre N,1990). Nucleic acids are regularly transferred from agarose gels to a membrane support by capillary action (Southern immunoprecipitation). Protein transfer evolved from DNA transfer (Southern) (Southern EM, 1975) and RNA transfer (Northern) (Alwine JC, and Kemp DJ, 1977). The term 'Western blot immunoprecipitation' was coined to describe (Burnette WN, 1981) this procedure in order to maintain the tradition of 'geographical' naming initiated by Southern's paper (Southern EM, 1975). The transferred proteins form an exact replica of the gel and have proved to be the first step in various experiments. The subsequent use of antibody probes directed against membrane-bound proteins (immunoprecipitation) revolutionised the field of immunology (Fig. 42).

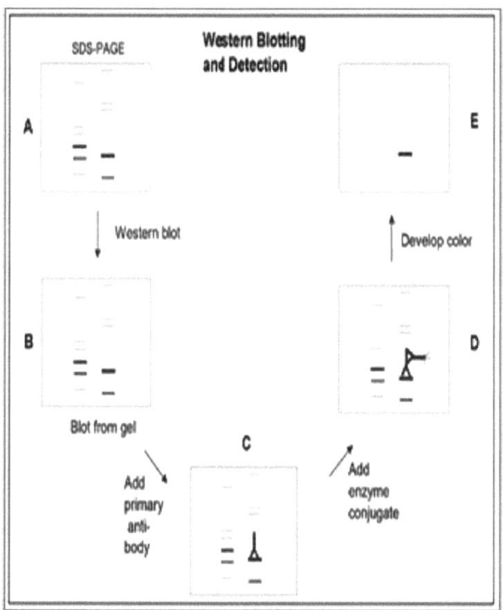

Figure 42:
Schematic representation of Western blotting and detection procedure (A) Unstained SDS PAGE gel prior to Western blot. The bands shown are hypothetical (B) Exact replica of SDS PAGE gel obtained as a blot following Western transfer. (C) Primary antibody binding to a specific band on the blot (D) Secondary antibody conjugated to an enzyme (alkaline phosphatase or horse radish peroxidase) binding to primary antibody (E) colour development of specific band (Reproduced from Ref 10with permission from Elsevier).

Dot blot refers to the analysis of proteins applied directly to the membrane rather than after transfer from a gel. The usefulness of high-resolution sodium dodecyl sulphate polyacrylamide gel electrophoresis (SDS-PAGE) was limited by the fact that proteins separated in the gel matrix were difficult to access with molecular probes, until the advent of protein immunoprecipitation. Protein transfer with subsequent immunodetection has found many applications in the fields of life sciences and biochemistry. This procedure (Towbin H, et al., 1979, LeGendre N,1990) is a powerful tool for detecting and characterising a multitude of proteins, particularly those that are not very abundant. It offers the following specific advantages: (a) wet membranes are flexible and easy to handle compared with gels, (b) proteins immobilised on the membrane are easily accessible to different ligands, (c) only a small quantity of reagents is needed for the blot analysis, (d) several replicates of a gel are possible, (e) prolonged storage of the blotted motifs before use becomes possible and (f) the same protein blot can be used for multiple successive analyses (Kost J, et al, 1994, Gershoni JM, 1988).

Protein immunoprecipitation has continued to evolve since its inception, and now the scientific community is faced with a multitude of means and methods of protein transfer (Kurien BT, and Scofield RH, 2006). Nevertheless, the sensitivity of Western blot immunoprecipitation depends on the efficiency of the transfer, the retention of the antigen during processing and the final detection/amplification system used. Results are compromised if there are deficiencies in any of these steps (Karey KP, and Sirbasku DA, 1989).

5.4 Polymerase chain reaction (PCR)

5.4.1 Definition of the PCR

This is a genetic technique that occurs in vitro and enables the enzymatic synthesis of large quantities (amplification) of a targeted region of DNA in an exponential manner. The DNA is synthesised in the same way as in vivo (in cells) using DNA polymerase (enzymes that cells use to replicate their DNA) (Mullis KB., 1990).

5.4.2 Principle of PCR

The polymerase chain reaction (PCR) is a powerful and widely used technical method that has greatly improved our ability to analyse genes. The genomic DNA present in cells contains several thousand genes. This makes it difficult to isolate and analyse any individual gene (Metzker, M. L. & Caskey, C. T., 2009).

5.4.3 Essential components of the reaction (PCR)

5.4.3.1 Model DNA

DNA template containing the patient's genomic DNA sample can be used in single or double stranded form. Closed circular DNA templates are amplified slightly less efficiently than linear DNAs. PCR requires only one copy of the target sequence as a template (Metzker, M. L. & Caskey, C. T., 2009).

5.4.3.2 Oligonucleotide primers

A pair of synthetic primers the oligonucleotides must be short, single-stranded and complementary to the opposite strands of the flanking regions of the fragment of interest. Standard reactions contain 0.1 - 0.5 µM of each primer, which is sufficient for 30 cycles of amplification of a 1 kb DNA segment. Higher primer concentrations favour poor priming and can lead to non-specific amplification. Oligonucleotide primers synthesised on an automated DNA synthesiser can be used for the PCR standard (Metzker, M. L. & Caskey, C. T., 2009).

5.4.3.3 Thermostable DNA polymerase

A thermostable DNA polymerase, which can withstand denaturation temperatures (94-95°C), is essential for catalysing the DNA-dependent synthesis model. Originally the Klenow fragment of Escherichia coli pol I was used (Saiki et al., 1985). This enzyme was inactivated at the high temperature required for the separation of two DNA strands and therefore had to be added freshly after each denaturation step. This obstacle was overcome with the introduction of thermostable DNA polymerase, Taq DNA polymerase isolated from the thermophilic bacterium Thermus aquatics. For a standard 25-50µl reaction, 0.5-0.25 units of Taq polymerase are used (Metzker, M. L. & Caskey, C. T., 2009).

5.4.3.4 Deoxynucleoside triphosphates (dNTP)

PCRs contain equimolar concentrations of...

a- dCTP, deoxycytidine triphosphate

b- dTTP, deoxythymidine triphosphate

c- dATP, deoxyadenosine triphosphate

d- dGTP, deoxyguanosine triphosphate (200-250 μM each).

These dNTPs are commercially available and supplied as pyrophosphate-free mixtures. The dNTPs should be stored at -200°C. During long-term storage, small amounts of water will evaporate and freeze to the walls of the vial. To minimise concentration change, vials should be centrifuged before use (Metzker, M. L. & Caskey, C. T., 2009).

11-3-5 Divalent cations

Free divalent cations are required for the activity of thermostable polymerases, and generally Mg+2 is used. Mg+2 binds to dNTPs and oligonucleotides. The molar concentration of cation must exceed the molar concentration of phosphate groups in the dNTPs and primers together. Hence the optimum concentration of cations must be calculated empirically for each reaction. A concentration of 1.5 mM Mg+2 is commonly used. An excess of Mg+2 will lead to an accumulation of non-specific substances in the amplification products and an insufficiency of Mg2+ will reduce the yield (Rittié L, Perbal B., 2008).

5.4.3.5 Buffer for maintaining pH

PH of the reaction mixture is adjusted to 8.3 - 8.8 at room temperature for the PCR standard (Innis MA, Gelfand DH., 2012).

5.4.3.6 Monovalent cations

50 mM KCl is used in standard PCR for amplification of DNA segments longer than 500 base pairs (Ruano G, Kidd KK., 1991).

5.4.3.7 Other

Some researchers have reported that the efficiency of the reaction is increased by the inclusion of 10% dimethylsulfoxide (DMSO) in the Taq polymerase buffer (Frackman S, et al., 1998, Farell EM, Alexandre G., 2012).

5.4.4 Amplification by Polymerase Chain Reaction (PCR)

One of the most crucial technical advances in molecular biology has been the use of specific probes or primers and the application of the principle of DNA replication to amplify sequences and assay very small fragments of DNA. If we know the nucleotide sequence of a gene, we can artificially produce oligonucleotides specific to the sequence. These primers can be used as a starting point for synthesising a copy of the gene using an enzyme called DNA polymerase, also known as Taq (from Thermus aquaticus, the micro-organism from which it is extracted) polymerase. This DNA polymerase resembles physiological polymerase, which is responsible for DNA duplication, but it is also thermostable, giving it a practical advantage.

PCR is a technique for amplifying DNA sequences almost infinitely, based on the principle of replication. As with replication, there is the action of a DNA polymerase which duplicates the sequence from primers. The two strands of DNA must then be artificially separated by heat (whereas during replication this operation is carried out under the action of a helicase).

The reaction (figs. 78 and 79) takes place in a small (and inexpensive) apparatus that is capable of changing temperature quickly, from 95°C, the temperature that dissociates the DNA, to 55°C, the temperature that allows the primers to hybridize, and finally to 72°C, the temperature that allows the polymerase to have its optimum action. The primers are homologous copies of the ends of the sequence to be copied, one is 5'-3', hybridises to the 3' end of the non-coding strand and will allow de novo synthesis, with radioactive substrates, of the coding strand, the other is 3'-5'. PCR is used to amplify sparse DNA sequences or, after reverse transcription, RNA sequences (RT-PCR).

The nucleotide sequence between the primers will be fully amplified, with the two primers indicating the limits of the process on either side (fig. 43). In this way, it is possible to amplify a sequence for which only the ends are known, while the structure of the middle portion is unknown. This possibility is the basis for the development of microsatellite analysis in genetics, as we will discuss later (Bernard Swynghedauw and Jean-Sébastien Silvestre, 2008).

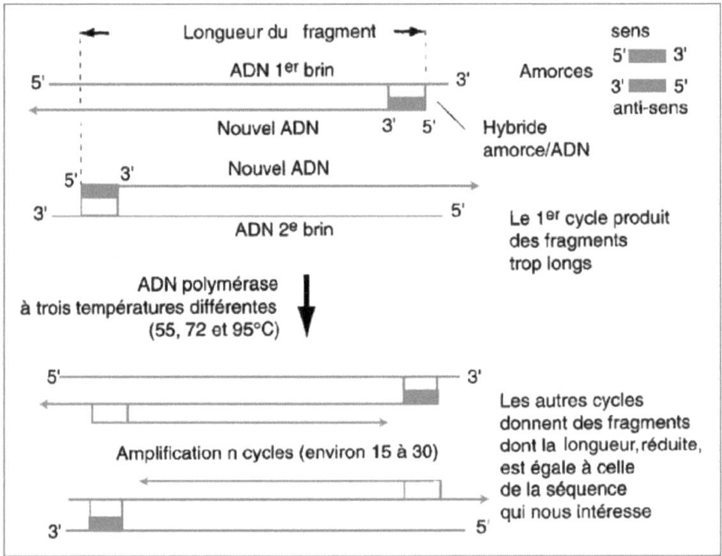

Figure 43 Principle of a PCR (Bernard Swynghedauw and Jean-Sébastien Silvestre, Polymerase Chain Reaction" is an artificial in vitro amplification using primers.
The structure of the primers defines the length of the sequence to be amplified. By multiplying the cycles, minute quantities of DNA can be amplified until they become visible in UV electrophoresis (Figure 44).

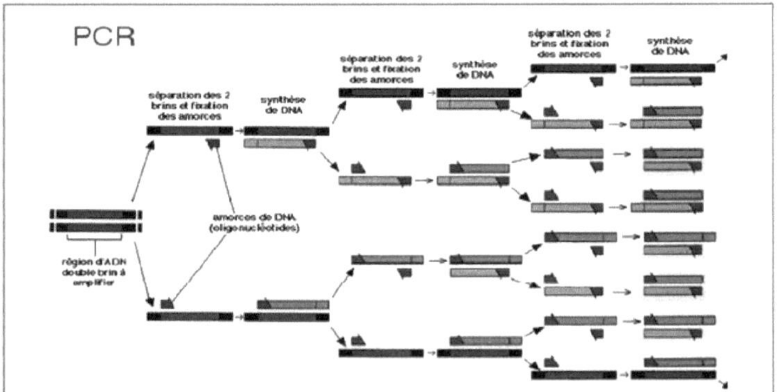

Fig 44. PCR (polymerase chain reaction) (C. Housset , A. Raisonnier, 2009 - 2010)

5.4.5 Types of PCR

In recent years, modifications or variants have been developed from the basic PCR method to improve performance and specificity, and to achieve amplification of other molecules of research interest such as RNA. Some of these variants are :

multiplex **a-PCR**, which simultaneously amplifies several DNA sequences (generally exonic sequences).

b-Nested PCR increases the specificity of the amplified product for a second PCR with new primers which hybridise to the fragment amplified in the first PCR.

Semi-quantitative **c-PCR**, which provides an approximation of the relative quantity of nucleic acids present in a sample.

d-RT-PCR, which generates RNA amplification by synthesis of cDNA (complementary DNA to the RNA), which is then amplified by PCR; and

Real-time e-PCR which performs relative quantification of nucleic acid copies obtained by PCR (Mohammad Ehtisham ,et al., 2016).

5.4.6 PCR applications

1. Detecting pathogens using primer pairs in clinical samples - All organisms have DNA sequences (rDNA) that code for ribosomal RNA. There are regions in the rDNA that vary between genera and species. These variable regions are amplified and then sequenced to determine the identity of the unknown organism.
2. Detection of viral pathogens and other micro-organisms that persist at low levels in infected cells and are difficult to identify by routine methods. Quantitative Real-Time can be used to detect viral genomes such as HIV or HPV.
3. Diagnosis of genetic disorders such as phenylketonuria, haemophilia, sickle cell anaemia and thalassaemia.
4. Identification of genetic mutations such as deletions, insertions and point mutations.
5. Screening of specific genes for unknown mutations.
6. Identification and analysis of mutations in eukaryotic DNA.
7. Genetic polymorphisms.
8. Gene expression
9. Forensic dentistry (Lo YD., 1998)

5.4.7 CONCLUSION

DNA evidence is also a powerful tool that has been used to finally prove the innocence of previously convicted individuals. This technique allows the specific in vitro amplification of very small numbers of a relevant DNA sequence to quantities that are suitable for study by conventional sequencing techniques. However, tests of the use of PCR in forensic analysis have amply demonstrated that these concerns are exaggerated, with even degraded samples giving reproducible and reliable results (Mohammad Ehtisham, et al., 2016) .

Chapter 4

Light and electron microscopy

6.1 Introduction

Microscopy has played an important role in determining the activity of cells, from Van Leeuwenhoek's (Ford 1989) very early appreciation of living animalcules with a simple microscope, to the details of cellular events with a variety of today's sophisticated imaging systems (Hell 2009). The current challenge is to look at living events with ever greater spatial and temporal resolution. The development of numerous transmitted light microscopy approaches, including techniques such as phase contrast, differential interference contrast (DIC) and polarised microscopy, have enhanced the inherent contrast of living specimens to make them more visible. However, the introduction of fluorescence microscopy, using a variety of fluorescent indicators (hereafter referred to as indicators) that can be tailored in terms of specificity for targets such as proteins, lipids or ions (Giepmans et al. 2006 ; Palmer and Tsien 2006) has perhaps been the most important step in enabling us to observe cellular physiology. Indeed, there seems to be no limit when it comes to designing innovative indicators using molecular approaches. But like all techniques, fluorescence microscopy is subject to practical physical limitations, the most important of which is resolution (Hell 2003). Optical microscopes can be categorised into bright field, fluorescence, phase contrast, quantitative phase and others such as scanning and transmission electron microscopes. In the following sections, more details will be given on each of the above microscopic modalities (Andreas Maier, et al., 2018).

6.2 Types of optical microscope

6.2.1 Basic optical microscope

a-Light propagates in the form of waves across ridges and valleys.

b-The amplitude of the peaks and troughs determines the luminosity of the light.

c-The number of times a complete wave occurs per unit time is called the frequency and the distance between two consecutive peaks is called the wavelength (λ) of the light.

d-The wavelength of the optical microscope between 400 and 700 nm makes up the visible spectrum.

e-The UV region is made up of wavelengths ranging from 100 to 385 nm.

f-Visualising any object directly through the human eye involves the incidence and reflection of light in the visual field.

g-Microscopes use daylight or light emitted by an incandescent bulb (Sameer Sheshrao Gajghate, 2016-2017).

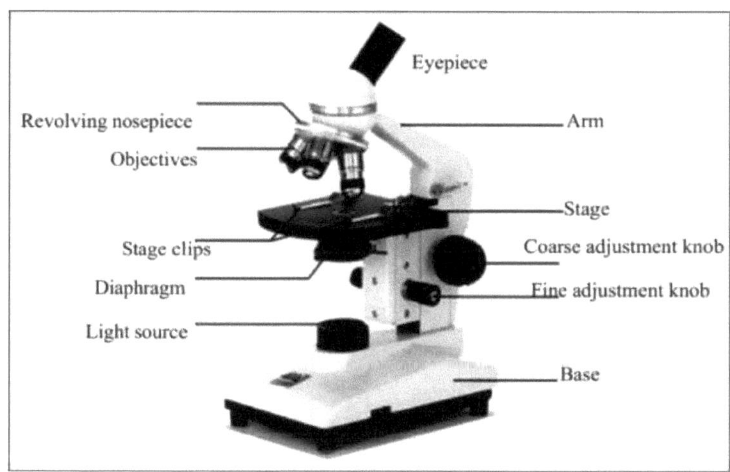

Fig. 45. Parts of the optical microscope

6.2.2 FLUORESCENCE MICROSCOPY

Fluorescence microscopy uses fluorescent dyes (fluorophores), which are molecules that absorb one wavelength of light (the excitation wavelength) and emit a second, longer wavelength of light (the emission wavelength). Most of the molecules in the cell are not very fluorescent, so the fluorescent labels to be imaged are usually introduced by the. This allows the markers to be targeted by the experimenter to the molecule(s) of interest, either by genetically encoding a fluorescent protein or by binding to a fluorescently labelled antibody. Several different fluorescent molecules can be distinguished simultaneously and detected in very low abundance (single molecules can be imaged), making it a very powerful technique. Fluorescence microscopy is generally performed using epifluorescence fig 46, in which the fluorescence excitation light illuminates the sample through the same objective used to detect the sample emission. A fluorescence filter cube separates the light by wavelength so that the emitted light can be imaged without interference from the excitation light (Murphy and Davidson, 2012). The two main techniques for introducing fluorescent markers into cells are immunofluorescence, in which fluorescently labelled antibodies that bind to specific proteins in cells are introduced, and genetic introduction of a fluorescent protein. In Immunofluorescence, the cells are first fixed to cross-link the proteins in the cell and then permeabilised to allow the antibodies access to the cell medium. Generally, the primary antibodies, which recognise the proteins of interest in the cell, are introduced first. Once the unbound antibodies have been removed, fluorescently labelled secondary antibodies, which bind to the primary antibodies, are added. This is known as indirect immunofluorescence and makes it easier to change primary antibodies, as they do not need to be labelled, and secondary antibodies generally have broad specificity (for example, a goat anti-mouse secondary that recognises all mouse immunoglobulin G). Genetic introduction of a fluorescent protein involves fusing a fluorescent protein to a target of interest, which is then either introduced into the cell's genome or expressed from a plasmid. This allows imaging of proteins in living cells and, if done by genomic introduction, means that the protein is expressed from its endogenous promoter at its endogenous level. For immunofluorescence and fluorescence imaging of proteins, it is straightforward to image four colours in a cell. Filter sets with excitation wavelengths of 405, 488, 561 and 640 nm are most commonly used. For immunofluorescence, typical dyes used are 4',6-diamidino-2-phenylindole nuclear dye, green dyes such as Alexa 488 or fluorescein, red dyes such as Cy3, rhodamine or Alexa 568 and dark red dyes such as Cy5. or Alexa 647. For fluorescent proteins, mTagBFP2, enhanced green fluorescent protein, mCherry/mRuby2/TagRFP-T and monomeric infrared fluorescent protein can be used together. Particularly in the 640 nm excitation channel, new fluorescent proteins are being developed rapidly and better combinations may emerge in the near future; for reviews of fluorescent proteins, see Day and Davidson (2009) and Dean and Palmer (2014) .

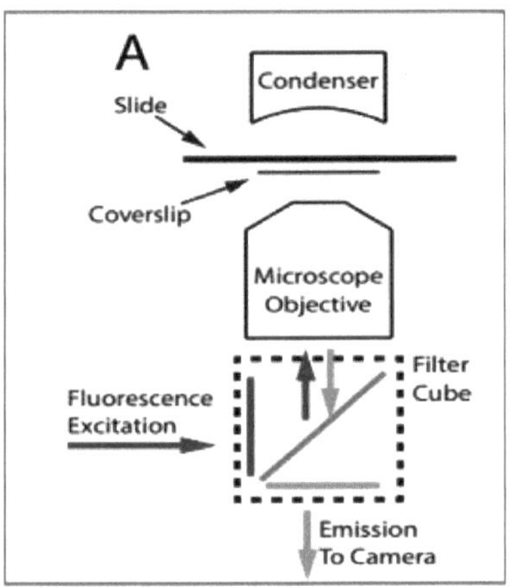

FIGURE 46: Schematic drawings of common microscopy techniques. (A) An inverted epifluorescence microscope. The sample is placed between the slide and the coverslip. The condenser lens provides illumination to view the light transmitted through the sample; the objective lens collects light from the sample and provides excitation light for fluorescence microscopy. The filter cube consists of an excitation filter (blue), an emission filter (green) and a dichroic mirror (grey). The excitation and emission filters respectively select the wavelengths that will illuminate the sample and be recorded on the camera, and the dichroic mirror reflects the excitation light back to the sample while transmitting the emission light to the camera. An upright microscope has the same

6.2.2.1 The disadvantages of fluorescence

Fluorescent probes are indeed very versatile, but suffer from three inherent disadvantages that need to be borne in mind during any imaging experiment.

Firstly, because they are self-luminous, fluorescent samples and slide preparations containing fluorescent cells will give a blurred image unless the sample is thin or you are viewing a stained monolayer of cells. High magnification objectives (due to their large numerical apertures) have an extremely limited depth of field, but a relatively large depth of focus. The depth of field is the axial depth of space on either side of the object plane within which the object can be moved without any detectable loss of sharpness in the image, and within which the features of the object appear sufficiently sharp in the image while the position of the image plane is maintained. The depth of field of a high NA microscope objective is typically <1 μm (see Table 7.5 in Sanderson , 2019). With very thick cells and tissues, a fluorescent signal emanating from outside the objective's depth of field will cause blurring in the image. The so-called optical section is used to remove the blur or prevent it from contributing to the image (Sanderson, J. 2020).

The second disadvantage of fluorophores is that they bleach. This is a consequence of the way electrons shuttle between atomic orbitals to give a fluorescent signal (Ishikawa - Ankerhold, Ankerhold and Drummen, 2012 ; Litchman and Conchello, 2005). The best ploy is to mount stained specimens in an anti-discolouration mounting medium (Bogdanov , Kudryavtseva and Lukyanov, 2012 ; Collins , 2006 ; Florijn , Slats, Tanke and Raap, 1993 ; Longin , Souchier, Ffrench and Bryon, 1993). and to minimise exposure to light. Keep slide preparations and samples refrigerated when not being imaged. It is easy to over-illuminate the sample, either with high-energy light in wide-field mode or with lasers in confocal mode, when focusing the sample, selecting an appropriate field of view or configuring imaging parameters (Sanderson, J. 2020).

The third disadvantage of fluorescent probes arises when two or more are used together to stain different targets in a single sample. If the excitation and emission spectral profiles of the fluorophores are close, particularly the emission profiles, then an emission signal can be detected in the channel of the other fluorophore(s). This is known as leakage, crosstalk or cross-emission (see Sanderson, 2019, for a discussion of these terms). To reduce the risk of leakage, select (if possible) fluorophores whose emission spectral profiles are widely separated. In addition, image the longest wavelength fluorophore first and, in particular, image fluorophores sequentially one after the other, rather than simultaneously. For a step-by-step procedure, refer to Basic Protocol 2: How to align the fluorescence bulb and set up the Köhler illumination for a wide-field fluorescence microscope (Sanderson, J. 2020).

6.2.3 Characteristics of bright field microscopy

As a general rule, the density and thickness of a sample vary in space. Consequently, points in the sample absorb light differently, i.e. the energy of the light after passing through the sample also varies in space. Figure 47 shows schematically how this fact can be used in a microscopic configuration.

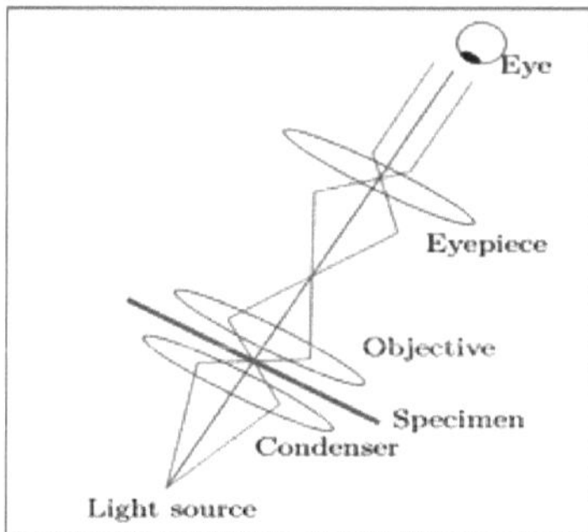

Fig 47: Scheme of a bright field

- light background
- low contrast

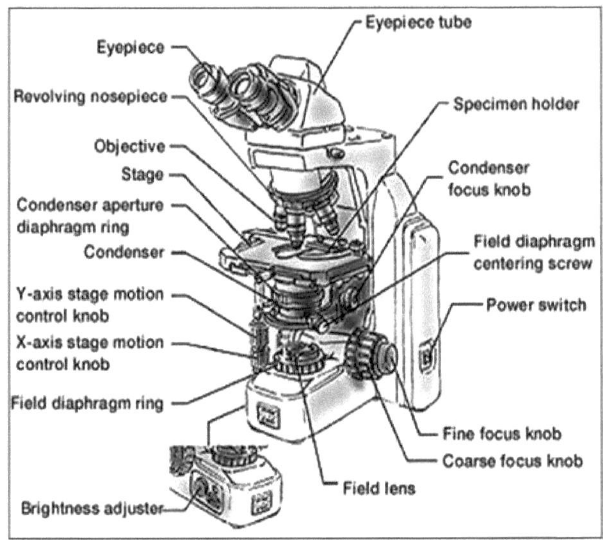

Fig. 48. Phase contrast microscopy [Gabriel Popescu, 2002]

The condenser shown in Figure 48 concentrates the light from a light source onto the sample. Information about the sample is encoded in the intensity of the light wave that reaches the lens. The background or the part of the scene that does not contain dense objects tends to be bright in the resulting image. This observer's impression gave the technique its name. The bright field configuration is the number one choice when minimising expenditure or implementation difficulties are major concerns. An example of a bright-field image of cells is shown in Figure 49 (Firas Mualla, et al., 2018).

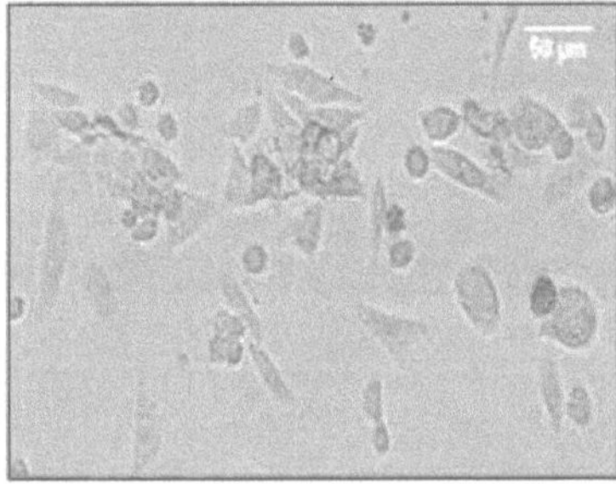

Figure 49: A microscopic image of a cell culture: The image was acquired using a Nikon Eclipse TE2000U microscope with a bright field objective of magnification 10× and NA = 0.3.

6.2.3.1 Somber field microscopy (fig 50)

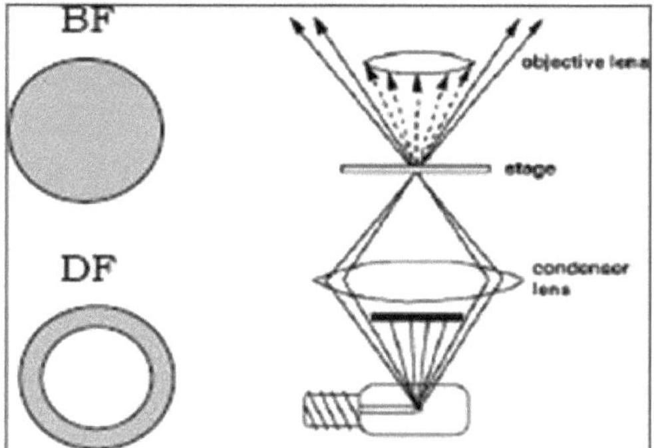

Fig. 50. Dark Field Microscopy [http://www.geog.ucl.ac.uk/~jhope/lab/micro23.stm]

a-an opaque disc is placed underneath
b-the condenser lens
c-scattered light
d-dark background
high e-contrast (structural details)

6.2.4 CONFOCAL MICROSCOPY

A major limitation of conventional epifluorescence microscopy is that the illuminating light excites fluorophores in a cone throughout the sample, and the detection camera cannot distinguish this blurred light from the light emitted from the focal plane of the sample. As a result, the focused information we are trying to image is obscured by blurred images of blurred regions of the sample. For samples that are not too thick and not too densely labelled with fluorophores, this blurred light is not a major problem. However, for thick and densely stained samples, or in cases where we wish to obtain well-resolved 3D images, this blurred light can mask valuable information. Numerous techniques have been developed to eliminate this blurred light. The most commonly used is confocal microscopy, in which the sample is illuminated by a laser beam focused on a single point in the focal plane of the sample (Figure 51). The light from this point is detected after passing through a pinhole, so that only the light emitted from the focal plane passes through the pinhole and is recorded on the detector. Light from the blurred planes is blocked by the pinhole, so the confocal only records light from the focal plane of the sample. Scanning mirrors are used to rasterise the laser spot on the sample, creating a point-by-point image (Inoué, 2006; Stelzer, 2006).

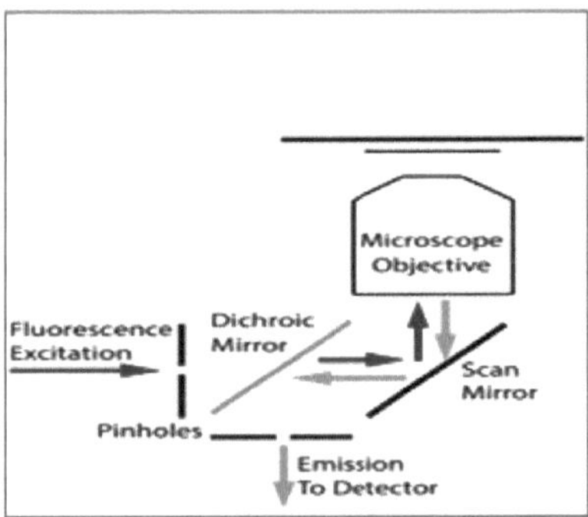

FIGURE 51: Schematic drawings of the common microscopes. A confocal microscope. Excitation and emission pinholes are imaged on the sample to define an illuminated point in the sample and detect light coming only from that point. The lenses that image the pinholes onto the sample have been omitted for simplicity. The scanning mirror scans the illuminated spot across the sample; Since the scanning mirror is in both the excitation and emission paths, the position of the spot detected from the sample is scanned in parallel with the excitation spot (Kurt Thorn, 2016).

A related technique is two-photon microscopy (Helmchen and Denk, 2005), which is mainly used for imaging very thick samples (> 200 µm), so it is not commonly used in cell biology. Because confocal laser scanning microscopes record a point-by-point image, they do not use cameras, but rather a point detector, which tends to be less sensitive than cameras. To overcome this limitation, systems that simultaneously scan several focal points on the sample and image the resulting emission on a camera have been designed. The most common of these is the spinning disc confocal, which uses a disc of pinholes that scan the sample in such a way that one revolution of the disc scans each point on the sample during a single exposure (Toomre and Pawley, 2006). Spinning-disk confocal microscopes combine ease of use, high speed (up to hundreds of images per second) and high sensitivity, making them widely used in cell biology. For samples thicker than ~30 µm, they reject blurred light less well than a confocal laser scanner, but this is not a limitation for imaging most tissue culture cells. Spinning-disk confocal microscopy is thought to be more respectful of living cells than wide-field or laser-scanning confocal microscopy, but definitive evidence is lacking. Spinning disc confocal microscopy is widely used for imaging protein and organelle dynamics in single cells, for example for imaging mitochondrial inheritance in yeast (Rafelski et al., 2012) or imaging microtubule dynamics in mammalian cells (Wittmann and WatermanStorer, 2005).
In the confocal microscope (Fig. 52), all blurred structures are removed during image formation.

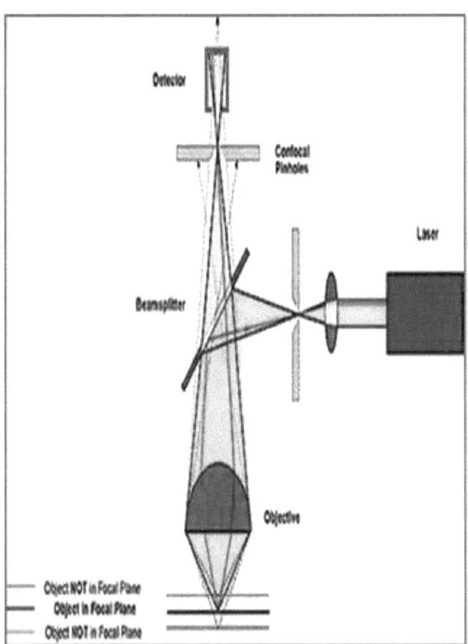

Fig. 52. Confocal microscope components [Philip D.

a-This is achieved by an arrangement of diaphragms which, at optically conjugate points in the path of the rays, act respectively as a source point and as a point detector.

b-The blurred rays are suppressed by the detection pinhole.

c-In addition to the wavelength of the light, the depth of the focal plane is determined by the aperture of the lens used and the diameter of the diaphragm.

6.2.4.1 Major improvements offered by a confocal microscope

The performance of a conventional microscope can be summarised as follows:

a-Light rays coming from outside the focal plane will not be recorded.

b-Defocusing does not create blur, but progressively cuts out parts of the object as they move away from the focal plane.

c-As a result, these parts become darker and eventually disappear. This feature is called optical sectioning.

d-Real three-dimensional data sets can be recorded.

e-Scanning the object in the x/y direction as well as in the z direction (along the optical axis) enables objects to be viewed from all sides.

f-Because of the small size of the point of light illuminating the focal plane, stray light is minimised.

g-With image processing, many slices can be superimposed, giving an extended-focus image that can only be obtained with conventional microscopy by reducing the aperture and thus sacrificing resolution (Sameer Sheshrao Gajghate, 2016-2017) .

6.2.4.2 Advantages and disadvantages of the optical microscope

6.2.4.2.1 Advantages

a-Direct imaging without sample pre-treatment, the only microscopy that allows true colour imaging.

b-Rapid and adaptable to all types of sampling systems, from gas to liquid and solid sample systems, of any shape or geometry.

c-Easy to integrate with digital camera systems for data storage and analysis.

6.2.4.2.2 Disadvantages

a-Low resolution, generally less than a micron or a few hundreds of nanometers, mainly due to the diffraction limit of light (**Sameer Sheshrao Gajghate, 2016-2017**) .

6.2.4.3 LIGHT SHEET MICROSCOPY

Finally, a revolution in microscopy has occurred with the development of light sheet microscopes (Weber and Huisken, 2011 ; Keller and Ahrens, 2015). These are microscopes that illuminate the sample from a plane orthogonal to the imaging plane (Figure 53). This eliminates the problem of blurred light because only light from the focal plane or very close to it is excited. This selective illumination also reduces the total light exposure of the sample, thereby reducing photobleaching and phototoxicity. If identical objectives are used to produce the light sheet and detect fluorescence, their roles can be reversed, with orthogonal views of the sample captured through both objectives. These images can then be merged to produce a 3D image of the sample with isotropic resolution, eliminating the poorer Z resolution of other forms of microscopy (Wu et al., 2013).

Alternatively, a sophisticated optical design can be used to produce microscopes capable of capturing high-resolution 3D images at high speed (Chen et al., 2014)

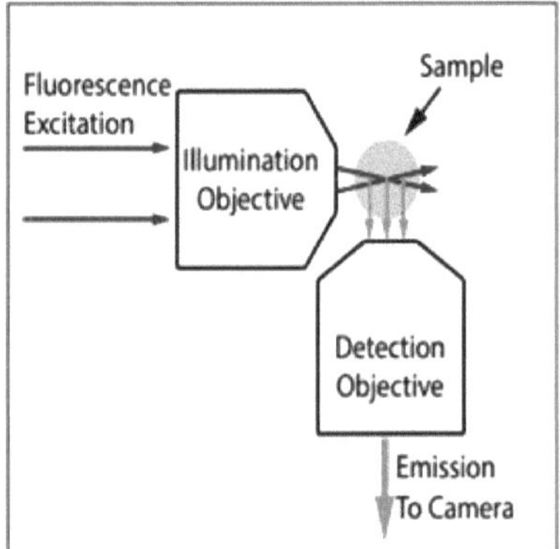

Fig 53: A light sheet microscope. An illumination objective, together with additional optics (not shown), are used to form a thin sheet of light that illuminates the sample. A detection objective images the light emitted by this sheet onto a camera (Kurt Thorn, 2016).

6.3 Electron microscopy

a- Electron microscopes are scientific instruments that use a beam of energy to examine objects on a very fine scale.

b-Electron microscopes use a beam of electrons instead of light

c-The object cannot be perceived directly by our eyes.

e-The image produced by electron microscopes is perceived by CRT or X-ray plates.

f- Electron microscopes were developed because of the limitations of optical microscopes, which are restricted by the physics of light.

g- By the early 1930s, this theoretical limit had been reached, and there was a scientific desire to see the finest details of the inner structure of organic cells (nucleus, mitochondria, etc.).
h- This required an increase in magnification of 10,000 times, which was not possible with existing optical microscopes (**Sameer Sheshrao Gajghate, 2016-2017**).

6.3.1 Summary of electron microscope components

6.3.1.1 The electronic optical column consists of :

a- An electron source to produce electrons
b- Magnetic lenses for beam coarsening
c- Magnetic coils to control and modify the beam
d- An aperture to define the beam, prevent electron dispersion, etc.

6.3.1.2 Vacuum systems consist of :

a- A chamber to maintain the vacuum, pumps to produce the vacuum
b- Valves to control vacuum, gauges to monitor vacuum

6.3.1.3 Signal detection and display consists of :

a- Detectors that collect the signal
b- Electronic components that produce an image from the signal.

6.3.2 Different types of electron microscope

6.3.2.1 Scanning electron microscopy (SEM) :

This electron microscopy-based technique determines the size, shape and morphology of the surface with direct visualisation of the nanoparticles. Scanning electron microscopy therefore offers several advantages in terms of morphological and dimensional analysis. However, they provide limited information on the actual size distribution and average of the population. During the SEM characterisation process, the nanoparticle solution must initially be converted into a dry powder. This dry powder is then mounted on a sample holder and coated with a conductive metal (e.g. gold) using a sputter coater. The entire sample is then scanned with a fine focused electron beam. The secondary electrons emitted by the sample surface determine the characteristics of the sample surface. This electron beam can often damage the polymer of the nanoparticles, which must withstand the vacuum.
The average size assessed by SEM is comparable to the results obtained by dynamic light scattering. In addition, these techniques are time-consuming, expensive and often require additional information on the size distribution (Ahmed Naji Al-Jamal, 2020).
a-It provides a valuable combination of high-resolution imaging, elemental analysis and, recently, crystallographic analysis.
b-It is used to inspect the topography of specimens at very high magnifications using equipment called a scanning electron microscope.
c-SEM magnifications can go up to more than 300,000 X, but most semiconductor manufacturing applications require magnifications of less than 3,000 X only.
d-It is often used to analyse cracks and fracture surfaces in matrices/cases, bond failures and physical defects on the surface of the matrix or case.
e-Identify crystalline compounds and determine the crystallographic orientations of microstructural features as small as 1 µm (recently developed capability - currently little used, but likely to become so) (**Sameer Sheshrao Gajghate, 2016-2017**).

6.3.2.1.1 Principle of operation of the scanning electron microscope

A-The main components of a typical SEM are the electron column, the scanning system, the detector(s), the display, the vacuum system and the electronic controls shown in Figure 54.

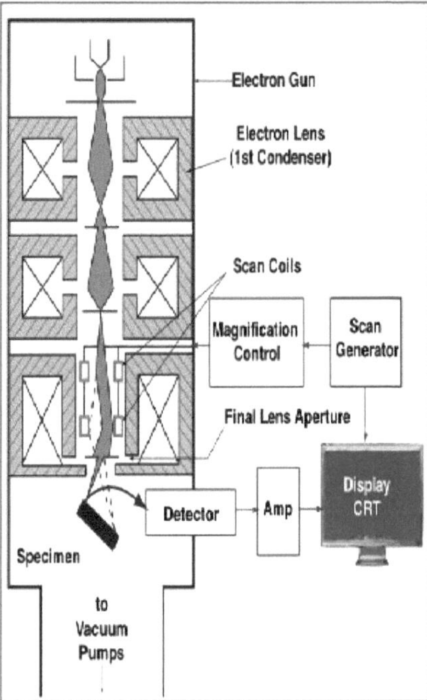

Fig. 54. SEM operating principle [M.T. Postek, et al., 1980].

B-The electron column consists of an electron gun and two or more electromagnetic lenses operating in a vacuum.

C-The electron gun generates free electrons and accelerates them to energies of between 1 and 40 keV in the SEM.

D-The purpose of electronic lenses is to create a small electronic probe focused on the sample.

E-Most SEMs can generate an electron beam on the surface of the sample with a spot size of less than 10 nm.

F-Max. sample size can be used up to 2.5×10^{n} -7 nm.

G-To produce images, the electron beam is focused on a thin probe.

H-It is scanned over the entire surface of the sample using scanning coils (Fig. 54).

I-With a tension accelerating voltage, the penetration penetration of the electron isgreater and the interaction volumeis greater.

j-As a result, the resolution spatial micrographs created from these signals will be reduced.

k-There will therefore be a brighter image because the noof backscattered electrons (BSE) will increase but the resolution will be worse.

L-For secondary electron (SE) imaging at voltages typical (say 15 keV), SEBs can penetrate the secondary electron detector and degrade resolution because they originate deep within the sample.

M-The complex interactions of the electrons in the beam with the atoms in the sample produce a wide variety of radiation.

N-In this case, knowledge of electron optics, beam-sample interactions, detection and visualisation processes is required for successful use of SEM power (**Sameer Sheshrao Gajghate, 2016-2017**).

6.3.2.1.2 SEM components

6.3.2.1.2.1 Electron column

A-The electron column is where the electron beam is generated in a vacuum (Fig. 55).

B-Focused on a small diameter, it is scanned over the surface of a sample by electromagnetic deflection coils.

C-The lower part of the column is called the sample chamber.

D-The secondary electron detector is located above the stage inside the sample chamber.

E-The samples are mounted and fixed on the stage, which is controlled by a goniometer.

F-The manual controls for the stage are located on the sample chamber and allow x-y-z movement, 360° rotation and 90° tilt (Fig. 55).

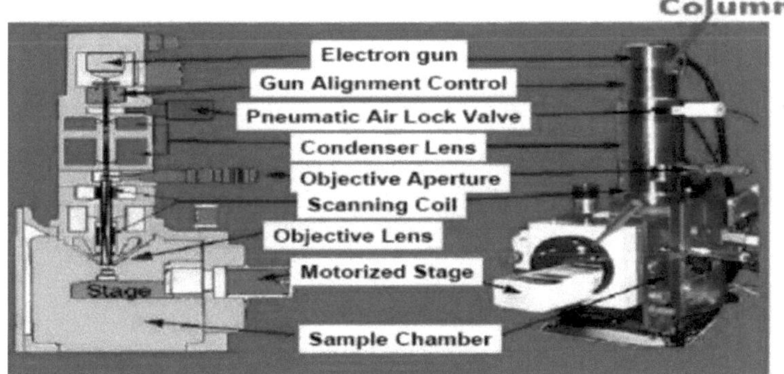

Fig. 55. SEM column [Doug, Holly & Oleg, "SEM Microscope].

6.3.2.1.2.2 Electron gun :

a-It is located at the top of the column.

b-Free electrons are generated by thermionic emission from a tungsten filament at ~2700K.

c-The filament is located inside the Wehnelt, which controls the number of electrons leaving the gun.

d-Electrons are mainly accelerated towards an anode that can be adjusted from 200 V to 30 kV (1 kV = 1,000 V).

6.3.2.1.2.3 Capacitor lenses :

a-Once the beam has passed through the anode, it is influenced by two condensing lenses that make the beam converge and pass through a focal point.

b-The electron beam is essentially focused up to 1,000 times its original size.

c-It is responsible for determining the intensity of the electron beam as it strikes the sample [MT. Postek et al., 1980].

6.3.2.1.2.4 Openings :

a-Depending on the microscope, one or more openings may be found.

b-The function of apertures is to reduce and exclude foreign electrons in lenses.

c-The final aperture of the lens below the scanning coils determines the diameter.

Or the size of the beam spot on the sample.

d-This will partly determine resolution and depth of field.

e-Decreasing the size of the spot will increase resolution and depth of field with a loss of luminosity [MT. Postek et al., 1980].

6.3.2.1.2.5 Scanning system :

a-The images are formed by tracing the electron beam through the sample using deflection coils inside the objective.

b-The stigmator or astigmatism corrector is located in the lens.

c-Use of a magnetic field to reduce electron beam aberrations.

d-The beam must have a circular cross-section when it strikes the specimen [MT. Postek et al., 1980, IM Watt, 1985].

e-Contrast with the predominant angular dependence of the secondary electron yield and edge effects.

6.3.2.1.2.6 Sample chamber :

a-The sample stage and controls are located in the lower part of the column.

b-Secondary electrons from the sample are attracted to the detector by a positive charge.

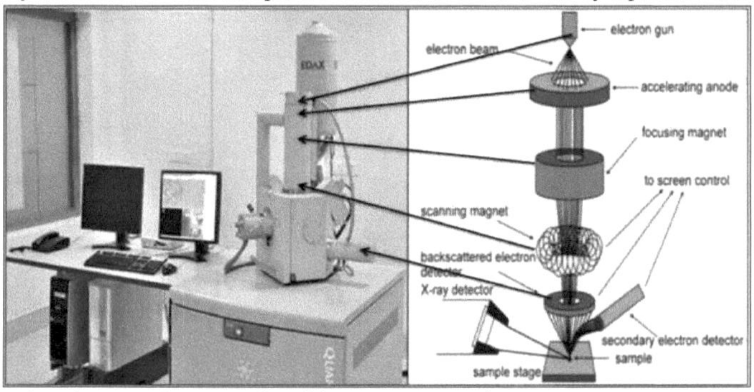

Fig 56: **Scanning electron microscope** (SEM) (M. Kannan, 2018).

6.3.2.1.2.7 Vacuum system

a-To provide a controlled electron beam, the column must be evacuated to a pressure of at least $5x10^{-5}$ Torr.

b-High vacuum pressure is required for a variety of reasons.

c-Firstly, the current flowing through the filament reaches temperatures of around 2700K [CE Lyman, 1990].

d-A filament will oxidise and burn in the presence of air at atmospheric pressure.

e-Secondly, to work properly, the column optics require a fairly clean and dust-free environment.

f-Thirdly, air particles and dust inside the column can interfere and block electrons [MT. Postek et al., 1980].

- In order to provide adequate vacuum pressure inside the column, a vacuum system consisting of two or more pumps is usually present.

6.3.2.1.2.8 Electron beam-sample interactions

- To achieve higher resolutions, an electron source is required instead of light as the illumination source.
- which allows resolutions of around 25 Angstroms.
- Not only does the use of electrons give better resolution but, due to the nature of the

interactions between the samples and the electron beams, there is a variety of signals.

- which can be used to provide information on the surface and near-surface characteristics of a sample (M.T. Postek, et al., 1980).

Table 2: Sample preparation for SEM (M. Kannan, 2018).

Step	Chemical	Temperatur e	Time	Repetitions
Primary fixing	2.5% glutaraldehyde in distilled water	Ambient or 0-4°C	2-4 hours or microwave	1
Wash	distilled water	Ambient or 0-4°C	30 minutes	3-5
Secondaries	1 to 4% osmium tetroxide in distilled water	Ambient or 0-4°C	2-4 hours	1
Wash	distilled water	Ambient or 0-4°C	30 minutes	3-5
Dehydration	25% ethanol	Ambient or 0-4°C	20 minutes	1
	50% ethanol		20 minutes	1
	70-75% ethanol		20 minutes	1
	90-95% ethanol		20 minutes	1
	100% ethanol		30 minutes	2

6.3.2.1.2.9 The dry critical point

Mount on a sample stub with silver paste or graphite

Spraying the biological sample with a gold/palladium alloy to make it conductive

Store the heels in a desiccator and visualise the outer surface with SEM (M. Kannan, 2018).

Fig 57: Scanning electron microscopy (SEM) micrographs (M. Kannan, 2018).

6.3.2.1.2.10 Applications of scanning electron microscopy

a-Topography: The characteristics of the surface of an object or "what it looks like", its texture; direct relationship between these characteristics and the properties of materials (hardness, reflectivity, etc.).

b-Morphology: The shape and size of the particles making up the object; direct relationship between these structures and the properties of the materials (ductility, strength, reactivity, etc.).

c-Composition: the elements and compounds that make up the object and their relative quantities; direct relationship between composition and material properties (melting point, reactivity, hardness, etc.).

d-Crystallographic information: How the atoms are arranged in the object; direct relationship between these arrangements and the properties of the materials (conductivity, electrical properties, resistance, etc.).

6.3.2.1.2.11 Advantages of SEM

> It provides detailed 3D and topographic imagery as well as versatile information gathered from a variety of sensors.
> This instrument works very fast.
> Modern SEMs allow data to be generated in digital form.
> Most SEM samples require minimal preparation.

6.3.2.1.2.12 Disadvantages of SEM

> SEM is expensive and bulky.
> Special training is required to operate a SEM.
> Sample preparation can give rise to artefacts.
> SEM is limited to solid samples.
> SEM carries a small risk of radiation exposure associated with electrons scattering below the sample surface **(M. Kannan, 2018).**

6.3.2.2 Transmission electron microscopy (TEM) :

Transmission electron microscopy (TEM) is the original form of electron microscopy and is analogous to the optical microscope. It can achieve a resolution of around 0.1 nm, a thousand times better than optical microscopy. The electron beam passes through the sample and analyses the internal structure of the sample in the form of images. The electron has a low penetration capacity and is absorbed in the thick sample. Consequently, the thickness of the sample should not exceed a few hundred angstroms (one angtron = $10^{à\ 10}$ m). However, sometimes slightly thicker samples are used in the high voltage electron microscope (M. Kannan, 2018) **.**

6.3.2.2.1 Instrument components (Sameer Sheshrao Gajghate, 2016-2017, M. Kannan, 2018, Sehasree Mohanta, 2021)

-Electron gun
-Condenser system (lenses and apertures to control illumination on the sample)
-Assembling the sample chamber
-Objective lens system

A. imaging lens - limits resolution ;
B. Aperture - controls imaging conditions

-Projector lens system (enlarges the image or diffraction pattern on the final screen)
-Image translation system (fluorescent screen and digital photographic unit)

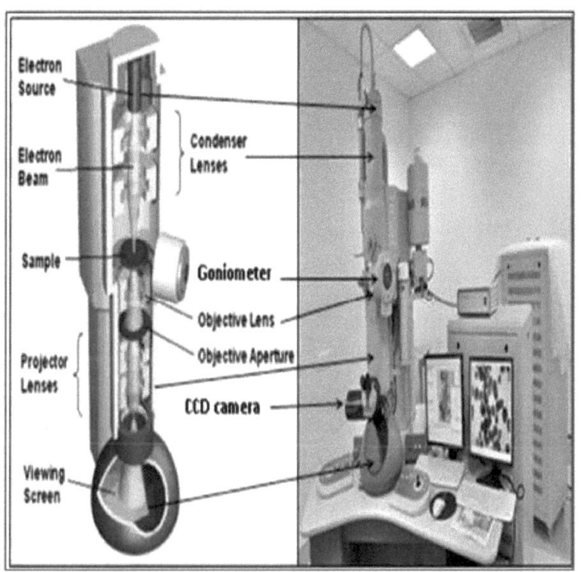

FIGURE 58**: Transmission electron microscopy components (M. Kannan, 2018)**

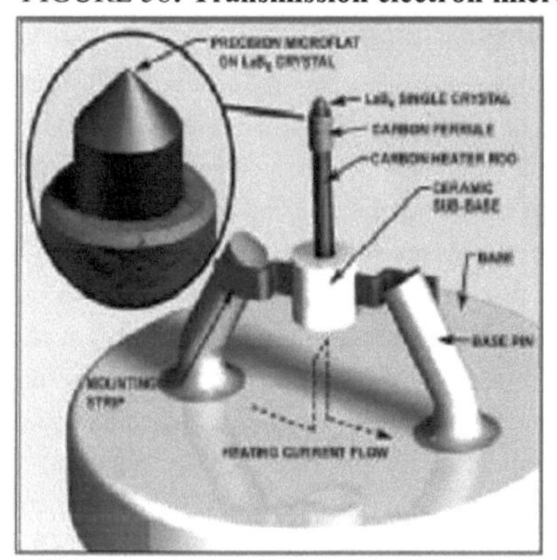

Fig 59: LB 6 electronic eraser source (**M. Kannan, 2018**).

Fig 60: Electromagnetic lens (**M. Kannan, 2018**).

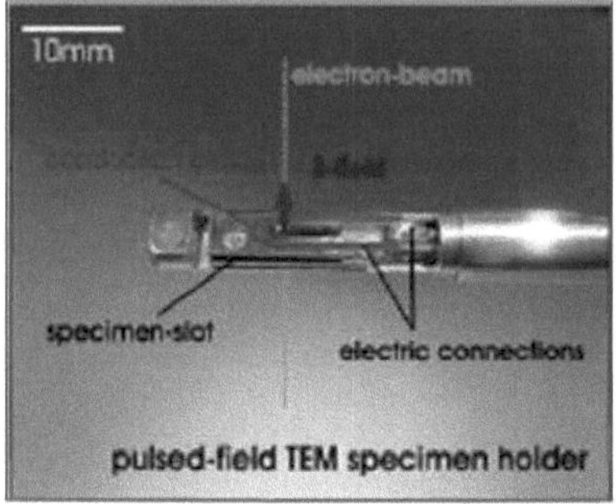

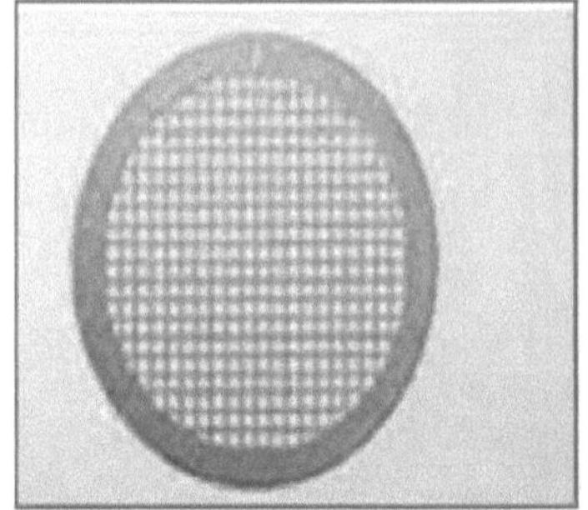

Fig 61: TEM sample holder (**M. Kannan, 2018**).
Fig 62: TEM copper grid mesh (**M. Kannan, 2018**).

6.3.2.2.2 Operating principles

TEM involves a high-voltage electron beam emitted from a tungsten filament (cathode) by electrical heating; the electron shaft is attracted to an anode (magnetic lens) and passes through an aperture. The beam passes through the aperture and then travels through an electromagnetic condenser, an objective, an intermediate and a projector lens. The focused electron beam is transmitted through a very fine sample (50 nm in size, semi-transparent to the electrons and carrying information about the structure of the sample) loaded onto a grid inserted in the path and manipulated by a goniometer. The absorbed part of the beam is scattered and transmitted through the objective aperture and projected by the projector objective after correction by intermediate lenses onto the fluorescence screen. The image is observed using optical binoculars attached to the viewing window. The spatial variation of the 'image' is then magnified by a series of magnetic lenses until it is recorded by striking a fluorescent screen or a light-sensitive sensor such as a CCD (charge-coupled device) camera mounted on the side or bottom of the image. photographic plate. TEM electron scattering rather than absorbance differences produces contrast in the image. Scattering results from an interaction between the atoms in the sample and the electrons in the light beam. Clouds of negatively charged electrons around atomic nuclei scatter the electrons by repelling them. These effects increase as the electrons move closer to the specimen atoms. Higher atomic number nuclei, as in heavy metal atoms such as lead and uranium, cause wider scattering. Transmission electron microscopes produce two-dimensional black and white images. TEM can easily resolve structures such as ribosomes, microtubules, microfilaments and large molecules such as proteins. Even images of individual heavy metal atoms have been produced under operating conditions.
Special (**M. Kannan, 2018**).

6.3.2.2.3 Biological sample preparation techniques for TEM (Dr M. Kannan, 2018).

a-Isolation of tissues
b-Fixation with glutaraldehyde, OsO4 and occasionally KMnO4
c-Encased in plastic resin
d-Ultramicrotomy
e-Post-colouring of thin sections

f-Photography

Table 4: Sample preparation

STEP	Chemical	Temperature	Time	Rehearsals
Fixing primer	2.5% glutaraldehyde in buffer	ambient or 0-4°C	2 to 4 hours or in the microwave	1
Wash	Tampon	ambient or 0-4°C	30 minutes	3-5
Secondaries	1-4% osmium tetroxide in buffer	ambient or 0-4°C	2-4 hours	1
Wash	buffer or distilled water	ambient or 0-4°C	30 minutes	3-5
en bloc (optional)***	0.5% uranyl acetate	0-4°C	during the night	1
Wash after *block* staining	distilled water	Ambient or 0-4°C	10-15 minutes	2
Dehydration	25% ethanol 50% ethanol 70-75% ethanol 90-95% ethanol 100% ethanol Transition solvent if the mounting resin is not miscible with ethanol	Ambient or 0-4°C	20 minutes 20 minutes 20 minutes 20 minutes 30 minutes	1 1 1 1 2
Infiltration	1 part resin/2 parts solvent 1 part resin/1 part solvent (optional) 2 parts resin/1 part solvent 100% resin	Ambient Ambient Ambient Ambient	1 hour - all night 1 hour - all night 1 hour - all night 1 hour	1 1 1 1
Integration	Place in 100% resin in a suitable container	1	1	11
Degassing (optional)	Place in a vacuum desiccator or vacuum oven	Ambient - 60°C	3-30 minutes	
Polymerisation	Place in a vacuum desiccator or vacuum oven	60-70°C	> 8 hours	
Ultramicrotomy	Cutting of the capsule Manufacture of glass knives Sectioning - Thick sections (size 500 nm) - optical microscopy - Thin sections (size 50 nm) -			

	TEM Staining sections - Thick sections - Light microscopy with toluidine blue staining - Thin sections (size 50 nm) - Negative staining with 1-2% phosphotungstic acid (PTA) or uranyl acetate or bacteria, Virus, bacteriophage and cell fragment samples Photograph -			
	Image capture from a thin section			

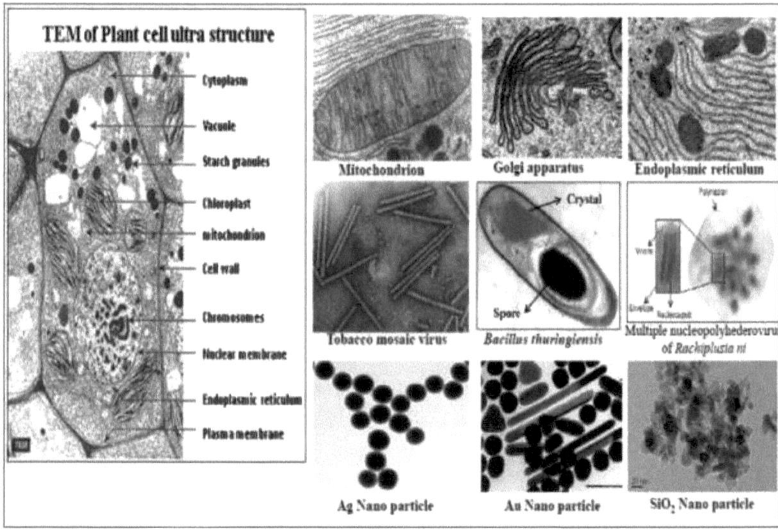

Fig 63: TEM micrograph of plant cell organelles (**M. Kannan, 2018)**

6.3.2.4 Application of TEM (M. Kannan, 2018, Ahmed Naji Al-Jamal, 2020, Sehasree Mohanta, 2021):

a-The transmission electron microscope is ideal for a number of different fields, including life sciences, nanotechnology, medical, biological and materials research, forensic analysis, gemology and metallurgy, as well as industry and education.

b-MET's provide topographical, morphological, compositional and crystalline information.

c-Images allow researchers to visualise samples at the molecular level, enabling structure and texture to be analysed.

d-This information is useful in the study of crystals and metals, but also has industrial applications.
e-TEMs can be used in the analysis and production of semiconductors, as well as in the manufacture of computer chips and silicon.
f-Technology companies are using TEM to identify defects, fractures and damage on micrometre-sized objects; this data can help solve problems and/or create a more durable and efficient product.
g-Colleges and universities can use MMT for research and studies.
h- Although electron microscopes require specialised training, students can assist teachers and learn TEM techniques.
I-Students will have the opportunity to observe a nanoscale world with incredible depth and detail.

6.3.2.2.5 Advantages and disadvantages of TEM (M. Kannan, 2018, Ahmed Naji Al-Jamal, 2020).

6.3.2.2.5.1 Advantages

a- TEMs offer the most powerful magnification, potentially over a million times or more
b-TEMs have a wide range of applications and can be used in a variety of scientific, educational and industrial fields.
c-METs provide information on the structure of elements and compounds
d-High-quality, detailed images
e-TEMs are able to provide information on surface characteristics, shape, size and structure.
f-They are easy to use with the right training

6.3.2.2.5.2 Disadvantages

a-Some disadvantages of electron microscopes include:
b-MEMTs are bulky and very expensive
c-Laborious sample preparation
d-Potential artefacts resulting from sample preparation
e-Processing and analysis require special training
f-Samples are limited to those that are transparent to electrons, able to tolerate the vacuum chamber and small enough to fit in the chamber.
g-EMTs require special housing and maintenance
h-Pictures are in black and white
I-Electron microscopes are sensitive to vibrations and electromagnetic fields and must be placed in an area that isolates them from possible exposure.
g-A transmission electron microscope requires constant maintenance, in particular maintaining the voltage, the currents to the electromagnetic coils and the cooling water.

6.3.2.2.6 Comparison between microscopes

Unlike the TEM and optical microscope, the SEM image is not formed by an objective lens. The image is formed sequentially by scanning the surface of the sample with an electron beam.

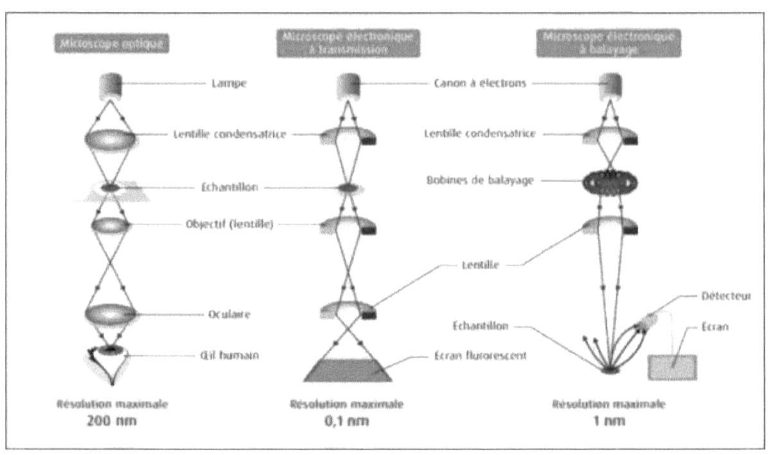

Fig 64: Comparison between microscopes

Table 1: Difference between light microscope and electron microscope (M. Kannan, 2018).

Sl. No.	Functionality	Optical microscope	Electron microscope
1	Electromagnetic Spectrum	Visible light, 400700 nm Visible colours	Electrons, application. 4 nm Monochromes
2	Maximum resolution Power	application. 200 nm	0.5 nm with very fine detail
3	Maximum magnification	x1000 tox1500	x500000
4	Source of radiation	Tungsten or quartz halogen lamp	High voltage (50kV) tungsten lamp, lanthanum hexaboride
5	Lenses	Glass	Electromagnetic
6	Interior	Filled with air	Vacuum
7	Focus screen	human hair (rètine), photographic film	Fluorescent screen (TV), Photographic film
8	Preparation of specimens	Living or dead temporary supports	Tissues must be dehydrated = dead
9	Fixing	Alcohol	OsO_4 or $KMnO_4$
Ten	Integration environment	Wax	Resin
11	Sectioning the specimen	Hand or microtome section < 20 μm slice Whole cells visible	Ultramicrotome sectioning Slices < 50 nm Visible parts of cells
12	Colour	Water-soluble dyes	Heavy metals

13	Taking charge of the sample	Glass blade	Copper grid

Table 5: Difference between TEM and SEM

Sl. No.	Functionality	TEM	MEB
1.	A beam of electrons	Wide and static beams	Beam focused on a precise point; the sample is one line scanned per line
2.	Necessary voltages	Much lower acceleration voltage; no need to penetrate the sample	SEM voltage varies from 60 to 300,000 volts
3.	Interaction of the electrical trons	The specimen must be very thin	Wide range of authorised specimens; simplifies sample preparation
4.	Imaging	Electrons must pass through and be transmitted by the sample	The necessary information is collected close to the surface of the specimen
5.	Image rendering	The transmitted electrons are collectively focused by the lens and magnified to create a real image.	The beam is scanned along the surface of the sample to build up the image.

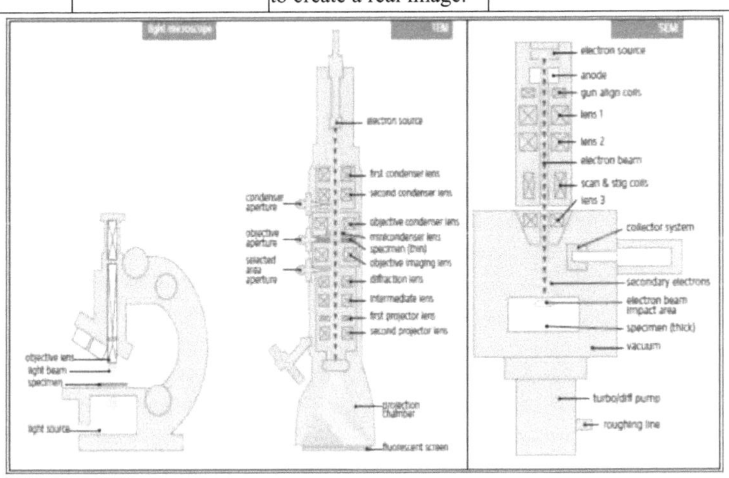

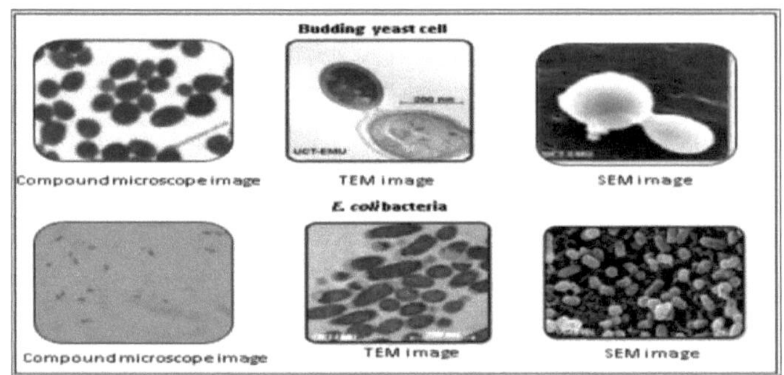

Figure 65: The **differences between TEM and SEM** (M. Kannan, 2018).

Chapter 5

General conclusion on laboratory analysis methods: In summary, the methods and techniques presented in this book highlight the crucial importance of laboratory analysis in various scientific fields. Whether in the context of biological and biochemical analyses or enzyme immunoassay procedures, these methods are essential tools for a better understanding of complex phenomena. The in-depth exploration of spectral methods, fractionation techniques, labelling methods and electron microscopy not only highlighted the versatility but also underlined the depth of laboratory analysis in uncovering the subtleties of biological and chemical systems.

Recommendations in the last section of the book:

Ongoing skills development: encourage readers to take part in ongoing training and skills development to keep abreast of advances in laboratory analysis techniques.

Collaborative research initiatives: highlight the importance of collaborative research projects bringing together experts from different disciplines to foster innovation and tackle complex scientific issues.

Regular updates and revisions: stress the importance of regularly updating the content of the book to reflect changes in methodologies and ensure that readers have access to the latest advances in laboratory analysis.

By incorporating these recommendations, the book aims to stimulate an ongoing quest for excellence in laboratory analysis and to contribute to the dynamic evolution of the scientific research landscape.

Bibliographical references

ADELINE. MT. X.P, WANG. C (1997) Evaluation of Taxoids from *TAXUS* sp. Crude Extracts by High Performance Liquid Chromatography.

Ahmed Naji Al-Jamal, Nanotechnology: fundamentals and applications, Al-Zawya Advertising Company, first edition, 2020, P 49- 50 , **578- 579** . (ISPN :) 3 - 326 - 20 - 9922 - 978

Alwine JC, and Kemp DJ (1977) Stark GR Method for detection of specific RNAs in agar gels by transfer to diazobenzyloxymethyl-paper and hybridization with DNA probes. Proc Natl Acad Sci USA 74, 5350-5354.

Andreas Maier, Stefan Steidl, Vincent Christlein, Joachim Hornegger (Eds.): Medical Imaging Systems, LNCS 11111, pp. 69-90, 2018. https://doi.org/10.1007/978-3-319-96520-8_5

Armstrong B. 2008. Antigen-antibody reactions. ISBT Science Series 3:21-32.https://doi.org/10.1111/j.1751-2824.2008.00185.x.

Bailey GS. 1996. Ouchterlony double immunodiffusion BT, p 749-752.in Walker JM (ed), The protein protocols handbook. Humana Press, Totowa, NJ.

Bernard Swynghedauw, BIOLOGY AND GENETICS MEMORIAL ASSISTANT MOLÉCULAIRES, © Dunod, Paris, 2008, 2000, © Nathan, 1994 for the 1st edition. ISBN 978-2-10-053798-3

BOATTO. G, CERRI. R, PAU. A, PALOMBA. M, PINTORE. G, GIOVANNA DENTI. M (1998). Monitoring of benzylpenicillin in ovine milk by HPLC. Journal of Pharmaceutical and Biomedical Analysis, Volume 17, August 1998, p 733-738.

Bogdanov, A. M., Kudryavtseva, E. I., & Lukyanov, K. A. (2012). Anti-fading media for live cell GFP imaging.Plos One, 7(12), e53004. doi: 10.1371/journal.pone.0053004

Boukhatem M.N. (2018). Aromatic and Medicinal Plants: the Scented Geranium. Botanical Description, Chemical Composition and Therapeutic Virtues. European University Publishing. ISBN : 6202277475

Bruneton, J. (1999). Essential oils. Pharmacognosy, phytochemistry, medicinal plants. Edition Tec & Doc, 3ème edition, Lavoisier, Paris, France.

Burgot G., Burgot J. L., Méthodes instrumentales d'analyse chimique et applications : Méthodes chromatographiques électrophorèses, méthodes spectrales et méthodes thermiques, 3ème Edition Tec & Doc Lavoisier, 2011, p.10, ISBN : 978-2-7430-1337-0.

Burnette WN (1981) "Western Blotting": electrophoretic transfer of proteins from sodium dodecyl sulfate-polyacrylamide gels to unmodified NC and radiographic detection with antibody and radioiodinated protein A. Anal Biochem. 112, 195-203.

Butcher DJ. Atomic Absorption Spectrometry | Interferences and Background Correction. Encyclopedia of Analytical Science. 2nd ed. Amsterdam, Netherlands: Western Carolina University, Cullowhee, NC, USA, Elsevier Ltd; 2005. pp. 157-163. DOI: 10.1016/b0-12-369397-7/00025-x

C. Housset , A. Raisonnier, Biologie Moléculaire, Objectifs au cours de Biochimie PAES, Université Pierre et Marie Curie, 2009 - 2010, Université Pierre et Marie Curie. P 192, 192, 195.

C.E. Lyman, D.E. Newbury, J.I. Goldstein, D.B. Williams, A.D. Romig, J.T . Armstrong, P . Echlin, C.E. Fiori, D.C. Joy, E. Lifshin and Klaus-Ruediger Peters, Scanning Electron Microscopy, X-Ray Microanalysis and Analytical Electron Microscopy: A Laboratory

Workbook, (Plenum Press. New York, N.Y ., 1990).
Calatayud JM, Icardo MC. Flow Injection Analysis, Clinical and Pharmaceutical Applications. Encyclopedia of Analytical Science. 2nd ed. Amsterdam, Netherlands: Elsevier Ltd; 2005. pp. 76-89. DOI: 10.1016/b0-12-369397-7/00159-x
Chemat, F., Abert-Vian, M., & Fernandez, X. (2013). Microwave-assisted Extraction for Bioactive Compounds. Food Engineering Series, Springer, New York, USA.
Chen B-C, Legant WR, Wang K, Shao L, Milkie DE, Davidson MW, Janetopoulos C, Wu XS, Hammer JA 3rd, Liu Z1, et al.(2014). Lattice light-sheet microscopy: imaging molecules to embryos at high spatiotemporal resolution. Science 346, 1257998.
CHEVIRON. N. A. ROUSSEAU (2000) Coumarin-SER-ASP-LYS-PRO-OH, A Fluorescent substrate for Determination of Angiotensin- coverting Enzyme Activity via High Performance Liquid Chromatography. Analytical Biochemitry, 280, 58-64.
Clément de Mecquenem, Marielle Drommi, Clémence Topart, Use of the rotary evaporator, Published on 20.06.18, https://culturesciences.chimie.ens.fr/thematiques/chimie-organic/methods-and-tools/using-the-rotary-evaporator
Collins, T. J. (2006). Mounting media and antifade reagents.Microscopy Today,14, 34-39. doi: 10. 1017/S1551929500055176.
CUQ. J-L (2007) Liquid Chromatography, page 4-48.
http://diffusiondessavoirs.uomlr.fr/balado/wpcontent/uploads/2008/01/chromato- liquid-2007.pdf (consulted on 05-02-2008).
Day RN, Davidson MW (2009). The fluorescent protein palette: tools for cellular imaging. Chem Soc Rev 38, 2887-2921.
De Caro CA, Claudia H. UV/VIS Spectrophotometry-Fundamentals and Applications. Schwerzenbach, Switzerland: Mettler-Toledo Publication No. ME-30256131; 2015
Dean KM, Palmer AE (2014). Advances in fluorescence labeling strategies for dynamic cellular imaging. Nat Chem Biol 10, 512-523.
Diehl B. Principles in NMR Spectroscopy. NMR Spectroscopy in Pharmaceutical Analysis. 1st ed. Elsevier Ltd; 2008. pp. 3-41
Doug, Holly & Oleg, "SEM Microscope".
DUMONTET. V, VAN HUNG, NGUYEN (2004) Cytotoxic Flavonoids and Pyrones from Cryptocarya obovate. Journal of Natural Products 67,858- 862.
Fabrice BRAY, La spectrométrie de masse haute résolution : Application à la FT-ICR bidimensionnelle et à la protéomique dans les domaines de l'archéologie et la paléontologie, Thesis by Fabrice Bray, Lille 1, 2017, P 2-3, Order number: 42555
Fan TWM, Lane AN. Applications of NMR spectroscopy to systems biochemistry. Progress in Nuclear Magnetic Resonance Spectroscopy. 2016;92-93:18-53
Farell EM, Alexandre G. Bovine serum albumin further enhances the effects of organic solvents on increased yield of polymerase chain reaction of GC-rich templates. BMC research notes. 2012 May 24;5(1):257.
Farhat, A. (2010). Microwave assisted vapo-diffusion: design, optimization and application. PhD thesis in Sciences (option: Process Sciences, Food Sciences), Université d'Avignon et des Pays de Vaucluse (France) & Ecole Nationale d'Ingénieurs de Gabès (Tunisia).
FEKETE, S., FEKETE, J., MOLNAR, I. GANZLER, K. (2009) Rapid high performance chromatography method development with high prediction accuracy, using 5 cm long narrow bore columns packed with sub-2 μm particles and design space computer modeling. *Journal of Chromatography A,* 1216, 7816-7823.

Ferhat, M. A., Meklati, B. Y., Smadja, J., & Chemat, F. (2006). An improved microwave Clevenger apparatus for distillation of essential oils from orange peel. Journal of Chromatography A, 1112(1), 121-126.

Ferhat, M. A., Meklati, B. Y., Visinoni, F., Vian, M. A., & Chemat, F. (2008). Solvent free microwave extraction of essential oils. Green chemistry in the teaching laboratory. Chimica Oggi, 21-23.

FIALKOV (A.B.) and AMIRAV (A.). - Cluster chemical ionization for improved confidence level in sample identification by gas chromatography/mass spectrometry. Rapid Commun. Mass Spectrom. 17, p. 1326-38 (2003).

Florijn,R.J.,Slats,J.,Tanke,H.J.,&Raap,A.K. (1993). Analysis of antifading reagents for fluorescence microscopy.Cytometry, 19(2), 177- 182. doi: 10.1002/cyto.990190213

Ford BJ. Antony van Leeuwenhoek-Microscopist and visionary scientist. *J Biol Education.* 1989;23:293-299

Frackman S, Kobs G, Simpson D, Storts D. Betaine and DMSO: enhancing agents for PCR. Promega notes. 1998 Feb;65(27-29):27-9.

Francis Rouessac, Annick Rouessac with the collaboration of Daniel Cruché, ANALYSE CHIMIQUE Méthodes et techniques instrumentales modernes, Cours et exercices corrigés, DUNOD 6eme edition, Dunod, Paris, 2004, Masson, Paris, 1992 for the 1ère edition ISBN 2 10 048425 7

Francis Rouessac, Annick Rouessac with the collaboration of Daniel Cruché, ANALYSE CHIMIQUE Méthodes et techniques instrumentales modernes, Cours et exercices corrigés, DUNOD 6eme edition, Dunod, Paris, 2004, Masson, Paris, 1992 for the 1ère edition ISBN 2 10 048425 7

Gabriel Popescu, 2002, "Chapter 4. Principles of Optical Imaging" Electrical and Computer Engineering, University of Illinois at Urbana-Champaign, Beckman Institute Quantitative Laboratory, Light Imaging. http://light.ece.uiuc.edu.

Gavahian, M., & Chu, Y. H. (2018). Ohmic accelerated steam distillation of essential oil from lavender in comparison with conventional steam distillation. Innovative Food Science & Emerging Technologies, 50, 34-41.

Germer TA, Zwinkels JC, Tsai BK. Theoretical concepts in spectrophotometric measurements. Experimental Methods in the Physical Sciences. 2014;**46**:11-66

Gershoni JM (1988) Protein blotting: a manual. Methods Biochem Anal. 33, 1-58. Review.

GHASHGHAIE. J, M DURANCEAU (2001) □ 13C of CO_2 respired in the dark in relation to □ 13C of leaf metabolites: comparison between Nicotiana *sylvestris andHelianthus annuus* under drought Plant, Cell and Environment, 24, 505- 515.

Giepmans BN, Adams SR, Ellisman MH, Tsien RY. The fluorescent toolbox for assessing protein location and function. *Science.* 2006; 312:217-224.

Golmakani, M. T., & Rezaei, K. (2008). Comparison of microwave-assisted hydrodistillation with the traditional hydro- distillation method in the extraction of essential oils from *Thymus vulgaris* L. Food Chemistry, 109(4), 925-930.

Gomes, P. B., Mata, V. G., & Rodrigues, A. E. (2007). Production of rose geranium oil using supercritical fluid extraction. Journal of Supercritical Fluids, 41(1), 50-60.

Günther H. NMR Spectroscopy; Basic Principles, Concepts and Applications In Chemistry. 3rd ed. John Wiley & Sons, Ltd, Chichester, U.K; 2013. p. 734

Hell SW. Microscopy and its focal switch. *Nat Methods.* 2009; 6:24-32.

Hell SW. Toward fluorescence nanoscopy. *NatBiotechnol.* 2003; 21:1347-1355.

Heimchen F, Denk W (2005). Deep tissue two-photon microscopy. Nat Methods 2, 932-940.
Hernandez Ochoa, L. R. (2005). Substitution of synthetic solvents and active ingredients by a combination (solvent/active ingredient) of plant origin. Doctoral thesis in Process Science (Agro-resource Sciences option), Institut National Polytechnique, Toulouse, France.
HOOIJSCHUUR (E.W.), KIENTZ (C.E.) and BRINKMAN (U.A.). - Analytical separation techniques for the determination of chemical warfare agents. J. Chromatogr. A, 982, p. 177200 (2002).
Hornbeck P. 2017. Double-immunodiffusion assay for detecting specific antibodies (Ouchterlony). Curr Protoc Immunol 116:2.3.1 2.3.4.https://doi.org/10.1002/0471142735.im0203s00.
Houot Robert. Separation between solids. In: Sciences
Géologiques. Bulletin, tome 46, n°1-4, 1993. Finely divided minerals. pp. 125-141; doi : https://doi.org/10.3406/sgeol.1993.1900
https://chem.libretexts.org/ Bookshelves/Physicaland TheoreticalChemistryTextbook
Hubert, R. (1992). Epices et aromates. Edition Tec & Doc, Lavoisier, France.
I.M. Watt, The Principles and Practice of Electron Microscopy, (Cambridge Univ. Press. Cambridge, England, 1985).
Innis MA, Gelfand DH. Optimization of PCRs. Innis MA, Gelfand DH, Sninsky JJ, White TJ, editors. Academic Press: San Diego, CA, USA; 2012 Dec 2.
Inoué S (2006). Foundations of confocal scanned imaging in light microscopy. In: Handbook of Biological Confocal Microscopy, ed. JB Pawley, New York: Springer, 1-19.
Ishikawa-Ankerhold, H. C., Ankerhold, R., & Drummen, G. P. C. (2012). Advanced fluorescence microscopy techniques-FRAP, FLIP, FLAP, FRET and FLIM. Molecules, 17(4), 4047-4132. doi: 10.3390/molecules17044047.
JACOB. V (2010) La chromatographie liquide haute performance. UIT de chimie de Grenoble. Date de mise en ligne : Wednesday 20 January 2010.
Kaloustian, J., & Hadji-Minaglou, F. (2012). The knowledge of essential oils:
Qualitology and aromatherapy: between science and tradition for a reasoned medical application. Collection Phytothérapie pratique, Springer-Verlag, Paris, France.
Karey KP, and Sirbasku DA (1989) Glutaraldehyde fixation increases retention of low molecular weight proteins (growth factors) transferred to nylon membranes for Western blot analysis. Anal Biochem. 178, 255-259.
Karoui R. Quality Control in Food Processing. Reference Module in Food Science, Encyclopedia of Food and Health. Amsterdam, Netherlands: Elsevier Ltd; 2016. pp. 567-572. DOI: 10.1016/b978-0-12-384947-2.00582-1
Keller PJ, Ahrens MB (2015). Visualizing whole-brain activity and development at the singlecell level using light-sheet microscopy. Neuron 85, 462-483.
Kindt T, Goldsby R, Osborne B. 2007. Kuby immunology. W.H. Freeman and Co, New York, NY.
Kost J, Liu L-S, Ferreira J, and Langer R (1994) Enhanced protein blotting from PhastGel media to membranes by irradiation of low-intensity. Anal Biochem. 216, 27-32.
Kurien BT, and Scofield RH (2006) Western blotting. Methods 38, 283-293.
Kurt Thorn, 2016 , A quick guide to light microscopy in cell biology, MBoC | TECHNICAL PERSPECTIVE, Volume 27, 219- 222. DOI:10.1091/mbc.E15-02-0088
Le BIHAN. J.Y, Le MASSON. J.P (2008) High performance liquid chromatography or H.P.L.C http://www.iut lannion.fr/LEMEN/Mpdoc/CHIMIE/Chimie1/CHROMATO.HTM.

(Consulted on 08-02-2008).
LeGendre N (1990). Immobilon-P transfer membrane: applications and utility in protein biochemical analysis. Biotechniques 9 (6 Suppl): 788-805. Review.
Leong YS, Ker PJ, Jamaludin MZ, Nomanbhay SM, Ismail A, Abdullah F, et al. UV-vis spectroscopy: A new approach for assessing the color index of transformer insulating oil. Sensors (Basel). 2018;18(7):2175
Leszczynska, D. (2007). Management de l'innovation dans l'industrie aromatique: Cas des PME de la région de Grasse. Editions l'Harmattan, Paris, France.
Lichtman, J. W., & Conchello, J. A. (2005). Fluorescence microscopy.Nature Methods, 2(12), 910-919. doi: 10.1038/nmeth817.
Ling Li X, Hu YJ, Mi R, Yun Li X, Qi Li P, Ouyang Y. Spectroscopic exploring the affinities, characteristics, and mode of binding interaction of curcumin with DNA. Molecular Biology Reports. 2013;40:4405-4413
Lo YD. Clinical applications of PCR. Springer Science & Business Media; 1998.
Longin, A., Souchier, C., Ffrench, M., & Bryon, P. A. (1993). Comparison of anti-fading agents used in fluorescence microscopy: Image analysis and laser confocal microscopy study.Journal of Histochemistry & Cytochemistry,41(12), 1833-1840.
Lucchesi, M. E. (2005). Microwave assisted solventless extraction: design and application to the extraction of essential oils. Doctoral thesis in Science (option: Chemistry), Faculty of Science and Technology, Université de la Réunion, France.
Lucchesi, M. E., Chemat, F., & Smadja, J. (2004). Solvent-free microwave extraction of essential oil from aromatic herbs: comparison with conventional hydro-distillation. Journal of Chromatography A, 1043(2), 323-327.
Luckie P., Hogg R. and Schaller R. (1980) - A Review of Two Fine Particle Processing Unit Operations -Classification and Mixing. In "Fine Particle Processing", P.
Somasundaran (Ed.), SME-AIME, chap. 10, p. 167-180.
M. Kannan, Transmission Electron Microscope - Principle, Components and Applications, in K S. SUBRAMANIAN, A Textbook on Fundamentals and Applications of Nanotechnology, DAYA Publishing House, 2018, P 81- 90, 93- 100. ISBN 9390384605, 9789390384600.
M.T . Postek, K.S. Howard, A.H. Johnson and K.L. McMichael, Scanning Electron Microscopy: A Student's Handbook, (Ladd Research Ind., Inc Williston, VT ., 1980).
M.T. Postek, K.S. Howard, A.H. Johnson and K.L. McMichael, Scanning Electron Microscopy: A Student's Handbook, (Ladd Research Ind., Inc. Williston, VT., 1980).
MAJORS. R. E, PRZYBYCIEL. M (2002) Columns for reversed-phase LC separations in highly aqueous mobile phases. *LC GC North America,* 20, 584-593.
Mancini G, Carbonnna AO, Haremans JF: Immunochemistry, 1965; 2: 235-254
MARCOZ. N (2003) Étude de stabilité des solutions injectables d'amiodarone. Diploma thesis in pharmaceutical analysis, University of Geneva, p 10-11.
MARRIOTT (P.J.), HAGLUND (P.) and ONG (R.C.). - A review of environmental toxicant analysis by using multidimensional gas chromatography and comprehensive GC. Clin. Chim. Acta, 328, p. 1-19 (2003).
Masango, P. (2005). Cleaner production of essential oils by steam distillation. Journal of Cleaner Production, 13(8), 833-839
Mendham J., Vogel A.I, Denny R.C., Toullec J., Barnes J., Barnes J.D., Mottet M., Tomas M.J.K., Analyse chimique quantitative de Vogel, Ed. De Boeck Université, 2005, p.231-314.
Metzker, M. L. & Caskey, C. T. (2009) Polymerase Chain Reaction (PCR). Encyclopedia of

Life Sciences (ELS), John Wiley & Sons, Ltd: Chichester.
Misra G. Fluorescence Spectroscopy. Data Processing Handbook for Complex Biological Data Sources. 1st ed. Academic Press; 2019. pp. 31-37
Mohammad Ehtisham , Firdous Wani , Iram Wani , Prabhjot Kaur , Sheeba Nissar, Polymerase Chain Reaction (PCR): Back to Basics, Indian Journal of Contemporary Dentistry, July-December 2016, Vol.4, No.2, 30-35. DOI: 10.5958/2320-5962.2016.00030.9
Mullis KB. The unusual origin of the polymerase chain reaction. Scientific American. 1990 Apr 1; 262(4):56-61.
Murali D, Venkatrao SV, Rambabu C. Spectrophotometric determination of etravirine in bulk and pharmaceutical formulations. American Journal of Analytical Chemistry. 2014; 5:77-82
Murphy DB, Davidson MW (2012). Fundamentals of Light Microscopy and Electronic Imaging, Hoboken, NJ: John Wiley & Sons.
Naira PEREZ VASQUEZ, CONTRIBUTION TO THE PROFILING OF URINARY ORGANIC ACIDS IN CHILDREN, DOCTORAL THESIS, UNIVERSITY OF PARIS-SUD, 2015
NIEMANN (H.B.), ATREYA (S.K.), BAUER (S.J.), BIEMANN (K.), BLOCK (B.), CARIGNAN (G.R.), DONAHUE (T.M.), FROST (R.L.), GAUTIER (D.), HABERMAN (J.A.), HARPOLD (D.), HUNTEN (D.M.), ISRAEL (G.), LUNINE (J.I.), MAUERSBERGER (K.), OWEN (T.C.), RAULIN (F.), RICHARDS (J.E.) and WAY (S.H.). - The Gas Chromatograph Mass Spectrometer for the Huygens Probe. Space Science Reviews, 104, p. 553-591 (2002).
OIV-Oeno 427-2010 Modified by OIV-COMEX 502-2012, Criteria for methods of quantification of potentially allergenic fining protein residues in wine, Method OIV- -MA-AS315-23, COMPENDIUM OF INTERNATIONAL METHODS OF ANALYSIS - OIV, Potentially allergenic fining protein residues in wine
OLIVEIRA. R, DE PIETRO. A, CASS. Q (2006) Quantification of cephalexin as residue levels in bovine milk by high-performance liquid chromatography with on-line sample cleanup. Talanta n° 71, p 1233-1238.
Olivero-Verbel, J., González-Cervera, T., Güette-Fernandez, J., Jaramillo-Colorado, B., & Stashenko, E. (2010). Chemical composition and antioxidant activity of essential oils isolated from Colombian plants. Revista Brasileira de Farmacognosia, 20(4), 568-574.
OSTROWSKI. T, JC MAURIZOT, MT ADELINE, JL FOURREY, P CLIVIO (2003) Sugar Conformational Effects on the Photochemistry of Thymidylyl (3'-5') thymidine. Journal of Organic Chemistry 68, 6502-6510.
Ouchterlony O. 1949. Antigen-antibody reactions in gels. Acta Pathol Microbiol Scand 26:507-515. https://doi.org/10.1111Zj .1699-0463.1949.tb00751.x.
Ouchterlony O. 1962. Diffusion-in-gel methods for immunological analysis. II. Prog Allergy 6:30-154.https://doi.org/10.1159/000313795.
Palmer AE, Tsien RY. Measuring calcium signaling using genetically targetable fluorescent indicators. *Nat Protoc.* 2006;1:1057-1065.
PANAIVA.L (2006) Chromatographic techniques for composite materials.
Eurocopter Conference, 1 June 2006, Marseille, France.
Passos MLC, Sarraguça MC, MLMFS S, Rao TP, Biju VM. Spectrophotometry | organic compounds. In: Reference Module in Chemistry, Molecular Sciences and Chemical
Engineering. Encyclopedia of Analytical Science (Third Edition). Amsterdam, Netherlands: Elsevier Ltd; 2019. pp. 236-243. DOI: 10.1016/b978-0-12-409547-2.14465-8

Pereira, C. G., & Meireles, M. A. A. (2010). Supercritical fluid extraction of bioactive compounds: fundamentals, applications and economic perspectives. Food and Bioprocess Technology, 3(3), 340-372.

Peterson, A., Machmudah, S., Roy, B. C., Goto, M., Sasaki, M., & Hirose, T. (2006). Extraction of essential oil from geranium (Pelargonium graveolens) with supercritical carbon dioxide. Journal of Chemical Technology & Biotechnology: International Research in Process, Environmental & Clean Technology, 81(2), 167-172.

European Pharmacopoeia. (2007). Council of Europe Directorate for the Quality of Medicines & Healthcare (EDQM), Strasbourg, France.

Philip D. Rack, on "Optical Microscopy", Dept. of Materials Science and Engineering University of Tennessee & lecture was generated by Professor James Fitz-Gerald at the University of Virginia.

polyacrylamide gels to NC sheets: procedure and applications. Proc Natl Acad Sci USA 76, 4350-4354.

Porstmann, T. and Kiessig S.T. Enzyme immunoassay techniques. An overview. *Journal of Immunological Methods.* 150 (1-2), 5-21 (1992).

POUPAT. C I. HOOK, F. GUERITTE. (2000), Neutral and basic taxoid contents in the needles of *TAXUS* species. Planta Medica, 66, 580-584.

PREMA, RAPURI (2003) HPLC: High-Performance Liquid Chromatography. Macmillian reference .USA, p165-167.

Raaman, N. (2006). Phytochemical techniques. New India Publishing, New Delhi, India.

Rafelski SM, Viana MP, Zhang Y, Chan Y-HM, Thorn KS, Yam P, Fung JC, Li H, Costa L da F, Marshall WF (2012). Mitochondrial network size scaling in budding yeast. Science 338, 822-824.

Rajpal S. Kashyap, Neha P. Agarwal, Nitin H. Chandak, Girdhar M. Taori, Saibal K. Biswas, Hemant J. Purohit, Hatim F. Daginawala, The application of the Mancini technique as a diagnostic test in the CSF of tuberculous meningitis patients, Med Sci Monit, 2002; 8(6): MT95-98, PMID: 12070446.

Rand RN. The role of spectrophotometric standards chemistry laboratory. Journal of Research of the National Bureau of Standards-A. Physics and Chemistry. 1972;**76**(5):499-508

REVILLARD With the collaboration of the Association des Enseignants d'Immunologie des Universités de Langue Française (ASSIM)IMMUNOLOGIE, 4th edition 2001- DeBoeck Université - ISBN2 - 8041 - 3805 - 4 , (Modified by Marie-Paule Lefranc, with the authorisation of Professor Jean-Pierre Revillard and DeBoeck Université (May 2003).

Rittié L, Perbal B. Enzymes used in molecular biology: a useful guide. Journal of cell communication and signaling. 2008 Jun 1;2(1-2): 25-45.

Rojas FS, CanoPavón JM. Spectrophotometry | Biochemical Applications. Encyclopedia of Analytical Science. 2nd ed. Amsterdam, Netherlands: Elsevier/The Lancet publishers, Elsevier Ltd; 2005. pp. 366-72

Rouessac F., Rouessac A., Brooks S., Chemical analysis: modern instrumentation methods and techniques, Ed. John Wiley and Sons, 2007, p.31.

Ruano G, Kidd KK. Coupled amplification and sequencing of genomic DNA. Proceedings of the National Academy of Sciences. 1991 Apr 1;88(7): 2815 9.

SALGHI. R (2004) Cours d'analyses physico-chimiques des denrées alimentaires II, GPEE, ENSA Agadir. Ecole Nationale des Sciences Appliquées d'Agadir p 7.

Sameer Sheshrao Gajghate, 2016- 2017, Introduction to Microscopy, Mechanical Engineering

Department National Institute of Technology Agartala
Sameer Sheshrao Gajghate, Introduction to Microscopy, Mechanical Engineering Department National Institute of Technology Agartala, 2016-2017
Sanderson, J. (2020). Fundamentals of microscopy. Current Protocols in Mouse Biology,10,e76. doi: 10.1002/cpmo.76
Sanderson, J. B. (2019).Understanding light microscopy. Chichester: John Wiley & Sons.
SANTOS (F.J.) and GALCERAN (M.T.). - Modern developments in gas chromatographymass spectrometry-based environmental analysis. J. Chromatogr. A, 1000, p. 125-51 (2003).
SAUNIER. C and GODIN. L (2013) Activité technologique en analyse biochimique, E.T.S.L paris, p. 5-50., http://ligodin.free.fr, date de consultation le 27/11/2010.
SCHELLINGER, A. P. & CARR, P. W. (2006) Isocratic and gradient elution chromatography: A comparison in terms of speed, retention reproducibility and quantitation. *Journal of Chromatography A,* 1109, 253-266.
SCIPPO. ML, MAGHUIN-ROGISTER. G (2006) Residues and contaminants in foodstuffs: 25 years of progress in their analysis. Biological screening methods. Ann. Méd. 150, 125-130.
Sehasree Mohanta, TRANSMISSION ELECTRON MICROSCOPY, July 2021. DOI:10.13140/RG.2.2.21280.71681
SNYDER. L. R, KIRKLAND. J. J. & GLAJCH. J. L (1988) *Practical HPLC Method Development.* New York: John Wiley & Sons.
Southern EM (1975). Detection of specific sequences among DNA fragments separated by gel electrophoresis. J Mol Biol. 98, 503-517.
Stelzer EHK (2006). The intermediate optical system of laser-scanning confocal microscopes. In: Handbook of Biological Confocal Microscopy, ed. JB Pawley, New York: Springer, 207220.
Stéphane Bouchonnet, Danielle Libong, Le couplage chromatographie en phasegazeuse-spectrométrie de masse, Département de Chimie, Laboratoire des Mécanismes Réactionnels, Ecole Polytechnique 91128 PALAISEAU Cedex, https://www.researchgate.net/publication/228763180
Suleyman Aydin. A short history, principles, and types of ELISA, and our laboratory experience with peptide/protein analyses using ELISA. *Peptides*, 72, 4-15 (2015).
SYAGE (J.A.), NIES (B.J.), EVANS (M.D.) and HANOLD (K.A.). - Field-portable, high-spedd GC/TOFMS.J. Am. Soc. Mass Spectrom, 12, p. 648-55 (2001).
THIERRY.B (2009) Cours de chromatographie liquide. Department of Chemistry, Université de la Réunion. http://www.uni-reunion.fr, Consultation date : 20/03/2012
Toomre D, Pawley JB (2006). Disk-scanning confocal microscopy. In: Handbook of Biological Confocal Microscopy, ed. JB Pawley, New York: Springer, 221-238.
Towbin H, Staehelin T, and Gordon J (1979) Electrophoretic transfer of proteins from
Tranchant J., Manuel pratique de chromatographie en phase gazeuse, 1995, Masson.
Trumbo TA, Schultz E, Borland MG, Pugh ME. Applied spectrophotometry: Analysis of a biochemical mixture. Biochemistry and Molecular Biology Education (Wiley Online Library). 2013;**41**(4):242-250
Upadhyay A, Upadhyay K, Nath N. Biophysical Chemistry (Principals and Techniques). Girgaon, Mumbai, India: Himalaya Pub House; 2009. ISBN: 1282802003 9781282802001
Ure AD, O'Brien JE, Dooley S. Quantitative NMR spectroscopy for the analysis of fuels: A case study of FACE Gasoline F. Energy Fuels. 2019;33(11):11741-11756

Verevkin SP, Zaitsau DH, Schick C, Heym F. Development of direct and indirect methods for the determination of vaporization enthalpies of extremely low-volatile compounds. Handbook of Thermal Analysis and Colorimetry. 2018;6:1-46. DOI: 10.1016/B978-0-444-64062-8.00015-2

Walton, N. N. J., & Brown, D. D. E. (1999). Chemicals from plants: perspectives on plant secondary products. World Scientific.

Wang, Z., Ding, L., Li, T., Zhou, X., Wang, L., Zhang, H., & He, H. (2006). Improved solvent- free microwave extraction of essential oil from dried *Cuminum cyminum* L. And *Zanthoxylum bungeanum* Maxim. Journal of Chromatography A, 1102(1), 11-17.

WANG. I. H, MOORMAN. R. & BURLESON. J (2003) Isocratic reversed-phase liquid chromatographic method for the simultaneous determination of (S)-methoprene, MGK264, piperonyl butoxide, sumithrin and permethrin in pesticide formulation. *Journal of Chromatography A,* 983, 145-152.

Weber M, Huisken J (2011). Light sheet microscopy for real-time developmental biology. Curr Opin Genet Dev 21, 566-572.

Wittmann T, Waterman-Storer CM (2005). Spatial regulation of CLASP affinity for microtubules by Rac1 and GSK3ßin migrating epithelial cells. J Cell Biol 169, 929-939.

Wu Y, Wawrzusin P, Senseney J, Fischer RS, Christensen R, Santella A, York AG, Winter PW, Waterman CM, Bao Z, et al.(2013). Spatially isotropic four-dimensional imaging with dual-view plane illumination microscopy. Nat Biotechnol 31, 1032-1038.

Yong Wang YN. Combination of UV-vis spectroscopy and chemometrics to understand protein-nanomaterial conjugate: A case study on human serum albumin and gold nanoparticles. Talanta. 2014;119:320-330

Printed by Books on Demand GmbH, Norderstedt / Germany